Marketing Telecommunications Services

New Approaches for a Changing Environment

For a complete listing of the *Artech House Telecommunications Library*,
turn to the back of this book.

Marketing Telecommunications Services

New Approaches for a Changing Environment

Karen G. Strouse

Artech House
Boston • London

Library of Congress Cataloging-in-Publication Data
Strouse, Karen G.
    Marketing telecommunications services : new approaches for a changing environment
    / Karen G. Strouse.
            p.   cm.—(Artech House telecommunication library)
        Includes bibliographical references and index
        ISBN 0-58053-015-X (alk. paper)
        1. Telecommunication—United States—Marketing.   2. Telecommunications
    policy—United States.   3. Competition—United States.   I. Title
    II. Series: Artech House telecommunications library.
    HE7775.S796   1999
        384'.0973—dc21   99-30838
                                                                                        CIP

British Library Cataloguing in Publication Data
Strouse, Karen G.
    Marketing telecommunications services : new approaches for a changing
    environment.—(Artech House telecommunications library)
    1. Telecommunication—United States—Marketing   2. Telecommunication
    policy—United States   I. Title
    384'.0688
    ISBN 1-58053-015-X

Cover design by Lynda Fishbourne

International Standard Book Number: 0-58053-015-X
Library of Congress Catalog Card Number: 99-30838
10 9 8 7 6 5 4 3 2 1

To my family

Contents

Preface

The U.S. telecommunications industry has been in the throes of deregulation for decades and is likely to remain in this transition well into the millennium. Meanwhile, many telecommunications carriers and regulators in other countries watch this evolution—with more than a passing interest—from their own positions on the path to privatization or from regulation to competition.

Few industries approach the size and reach of the telecommunications industry, yet its marketing practices have been only informally documented. The sound bites of telecommunications industry analysts and senior management regularly recognize that marketing differentiates the winners from the losers, but such analysis stops short of detail. *Marketing Telecommunications Services: New Approaches for a Changing Environment* aims to fill that void by synopsizing the marketing activities already in progress and predicting what the marketing environment will look like once a truly competitive telecommunications marketplace is implemented.

The book is intended for use as a handbook rather than as an industry analysis. Each chapter covers a distinct area of concern for marketers. All chapters except the first end with a self-assessment section to help readers begin the essential process of evaluating their present market positions and market readiness. Answers to the questions will be different for each telecommunications service provider, depending on broad issues of corporate direction and each provider's own history and capabilities. Service providers that have begun the transformation have learned that the journey is complex, energy-consuming, and difficult.

The telecommunications industry sits on a shifting foundation of deregulation, technology, and the individual creativity of its participants. As a result, it would be impossible for a telecommunications book to remain both accurate and timely from the completion of the manuscript to its final publication. Nevertheless, I have tried to include useful references and anecdotes on the current telecommunications environment to make a point. Some of these will no longer be valid—even before the book is printed. It is my hope that the marketing value of the examples will last longer than the company names, pricing structures, or the snapshot of the regulatory environment each provides.

I would like to thank the many colleagues, including coworkers and clients, who have helped me reorient my thinking from a monopoly outlook to a more practical competitive one. In addition, I would like to express my appreciation to the manuscript reviewers, who each provided constructive feedback, and to the team at Artech House for its constant support during the development process.

Part I

Introduction

1

The Lessons of History

The importance of marketing

For nearly a century, marketing in the telecommunications industry was indirect by most standards. Marketing largely consisted of rate-making and corporate communications activities. Marketing entailed communicating with federal and local regulators, consumer groups, and other stakeholders and interacting very rarely with individual customers. Sales activities were moderately passive. Most customers could not tailor a communications service to their needs, negotiate a better price, or use an alternative provider for services. The product itself, telephone service, was essential but unglamorous.

Marketing and selling skills essential to other industries were in short supply because they were not required to conduct business. The customer focus that is now heralded by most telecommunications service providers was virtually unknown. In all fairness, the concept of a customer focus did not affect most industries until the approach of the millennium. The current intersection of a customer focus against the backdrop of deregulation simply underscores its importance.

Telecommunications companies did not ignore their individual customers completely, but an absence of competition made their interactions with customers less meaningful. Marketers conducted research to determine the level of customer satisfaction with services for which customers had no choice of providers. Surveys asked about customers' willingness to pay for enhanced services they had never before seen or used and for which there would not be substitutes. Customers routinely reported that rates were too high for local exchange services that were paradoxically provided well below their costs. They generally reported satisfaction with long-distance prices that were often several multiples of what they would be without subsidies.

One reasonable conclusion to draw is that customers yearned for choice. Deregulation of all telecommunications services will offer choice —and much more. For providers, deregulation offers uncertainty, opportunity, and colossal change. Much of the change is in the marketing function, paramount in importance yet relatively untested for the incumbent providers. Their leading competitors are well-funded and eager to delve into the telecommunications services market. For some, marketing is already a core competency.

At the time airlines were deregulated, some industry giants, unable to transform their business practices and marketing capabilities, unthinkably perished. As the small or new companies disappeared, it was not especially newsworthy. When eminent Pan American Airlines did not survive deregulation, other industries needed to take note. Five years after local deregulation, telecommunications industry icons will disappear, and new leaders will take their places, largely due to their marketing strategies, skills, and execution.

Services marketing

It is tempting to assume that the marketing of services has requirements identical to those of marketing physical goods. However, there are four major differences between them: intangibility, inseparability, variability, and perishability.

Services are *intangible*. Because most services cannot be touched or used before their acquisition, it is difficult for customers to judge what they are buying before committing to the purchase. Services cannot be inspected, demonstrated, or packaged. Customers perceive more risk in service purchases than in their purchases of physical goods, creating selling barriers. Telecommunications providers have worked to overcome these limitations for competitive advantage through retail stores and packaging and by using symbols to substitute for merchandise because the merchandise itself cannot be seen or held. Telecommunications service providers will need to create tangible images for their services. Sprint's famous "pin drop" image provides a physical representation of its network quality. Wireless providers have been less successful by using animated mascots and animals in their advertising.

Services cannot be protected for years by patent. As telecommunications competition becomes more intense, this inability to protect new services by patent could result in copycat services becoming available to customers as soon as competitors can create them. Service providers will be unable to sustain high profits on new services after competitors develop substitutes. Moreover, information technologies have a shorter life cycle than their nontechnical counterparts. Accordingly, telecommunications providers will need to decide whether and when technology leadership is worth its cost.

In general, customers find it difficult to assess the value they place upon services of any kind. Most customers cannot imagine how the telecommunications services they use are produced or what it cost the provider to produce them. This customer inability, along with the high level of fixed cost that is generally recovered over time and a legacy of social objectives distorting the price structure, will make it especially challenging to price services in a competitive market. Offering some services well above costs and others below them complicates the expectations customers will bring to the marketplace. The fact that telecommunications is such a necessity for most people contributes to customer uncertainty.

Services are also *inseparable*. Services are produced and consumed at the same time. When a customer picks up a telephone and receives no dial tone, the customer is unimpressed that this failure is the first on the line in 10 years. Telecommunications customers expect services to be available—at a high-quality level—any time they are summoned. Other than historical experience, the service provider has few tools to forecast the moment-by-moment demand for the service and little time to reinforce the infrastructure if a service peak is imminent.

Similarly, services are *variable*. To the customer, the service and the provider are the same. The quality of the customer's experience is dependent on factors not under the control of the telecommunications provider, including representatives of the service provider, the customer, and the service environment. Telecommunications providers can reduce variability by automation, standardizing procedures, conducting intensive employee training, and strengthening the brand image.

The variability of telecommunications means that some customers can affect the quality of service for other customers. At any given time, the bandwidth available to cable modem users depends on the usage by other users. Providers can do little to control peaks or valleys of usage on the local and long-distance public network. Internet service providers (ISPs) do not control when users log on and consume their servers and network resources, creating excellent service for some customers and occasionally terrible service for others.

In addition, services are *perishable*. Service infrastructures are shared, designed to provide a particular level of service capability at any given time. The minute of network usage not sold right now is lost forever. Discounted pricing on weekends and evenings for long-distance and wireless services are intended to balance the traffic load on the network, which is engineered for a higher amount of traffic. Similarly, lost sales occur during times that the network is overburdened. The caller who generates the "all circuits busy" signal could decide not to make the call at all. Some telecommunications providers try to balance the demand on scarce teleconferencing or videoconferencing resources by scheduling their use in advance. Scheduling ahead conveniences the service provider. However, the provider that offers an unscheduled service creates a competitive differentiating factor.

Is competition here yet?

Regulators and others have struggled with metrics to establish whether each telecommunications market is indeed competitive, and that question will not be answered soon. A rule of thumb holding that competition is present when no provider holds more than half the market share at least establishes a baseline for discussion. By that standard, all markets except local service are competitive in the aggregate, and no local service market, including the business market, is competitive yet anywhere.

Competition is intense in the long-distance market. After the divestiture of its Bell operating companies in 1984, AT&T lost 30% of its market share in long-distance service, from 68.3% in 1984 to 40% in 1997 [1]. This drop occurred while regional Bell operating companies (RBOCs) were still precluded from entering the interexchange market. Most customers leaving the venerable long-distance carrier did not join AT&T's most well-known rivals, MCI (now MCI WorldCom) and Sprint, which both lost share during the same period, although not as much as AT&T did. The market winners were resellers, service providers with superior marketing capabilities and no network infrastructure, whose share increased significantly in the same period. Gaining share and holding onto it are two separate problems for carriers. Most industry analysts estimate long-distance churn, the percentage of customers who leave for competitors, at 30% or more.

In the wireless market, growth is strong, and competition is thriving. The wireless industry continues to grow at about 45–50% per year. The market entry of carriers using personal communications services (PCS) and other wireless technologies, coupled with cost reductions and other advances such as digital services, has enabled more wireless customers to choose from a larger variety of providers. Average prices for PCS services are 22% lower than digital cellular across the United States [2].

Major markets are intensely contested and constantly shifting. The Yankee Group reported in its 1998 survey that carrier-to-carrier churn among wireless customers rose from 7% to 10% in one year. Nearly one quarter of wireless customers have switched carriers at least once since they first became subscribers. Loyalty is weak; one of the most popular reasons for churning is simply that the service contract is over. Decision Resources estimates 1998 cellular churn at 2% per month [3]. The market researcher also calculates PCS network churn rates at double that of cellular for many operators and expects churn to increase to the level of long-distance services.

Local markets are slower to become competitive, but they are moving in an expected direction. Long-distance services, priced well above the cost of providing them, attracted competitors immediately. Wireless services, launched without much of a market history and with a backlog of demand, had few barriers to all-out competition. Local service presents more challenges to full competition. First, most local customers pay less for service than its cost to the provider. Second, connecting each customer's location to the public network requires a considerable investment in facilities. New competitors are reluctant to invest in facilities for individual customers, especially in the face of anticipated high churn. Resale, the alternative available to competitors, has lower cost but uncertain profitability.

Based on experiences in other competitive markets, incumbent local service providers stand to lose 15–30% of their customers each year when strong competitors arrive in their markets [4]. In exchange, they are expected to gain about 15–20% of the interexchange market from which the RBOCs are currently prohibited [5]. The Yankee Group predicted that by 2000, incumbent LECs would remain in control of 88.9% of the local exchange market, while interexchange carriers would hold 7.5%, and other competitive local exchange carriers (CLECs) a share of 3.6% [6]. Wireless, used as a substitute for wireline service, could take one-half of the market from incumbent providers by 2005 [7]. These predictions, while aggressive in their magnitude, are not as bullish about the timing of significant competition. The unwillingness to post short-term predictions reveals the widespread belief that local markets will be slow to become competitive. Three years after the signing of the Telecommunications Act deregulating the industry, local competition in consumer markets is barely detectable.

CLECs are penetrating selected markets. CLEC revenues are expected to double every year for a 15% market share before the millennium [8]. Three-quarters of U.S. businesses had a choice of local service providers by 1998, according to The Strategis Group. The market researcher also estimates that CLEC business market share will rise from 5% in 1998 to about 25% in 2003. Nevertheless, 80% of business customers surveyed reported that competition was having little or no effect on their corporate networks [9]. More than 90% of those surveyed said that having a choice of local carriers was important to them.

Monopoly's slow demise

Most observers would state that the telecommunications industry is undergoing deregulation. It would be more accurate to declare instead that the industry is under-deregulated. Regulators, long-distance providers, large customers, and local providers all insist that they expect to gain from deregulation. While each of the industry participants blames the others for the delays, with some justification, ghosts from the monopoly past are also accountable.

The transition to deregulation is complicated by the industry's social policy-led pricing. Universal service, the concept that basic local connections should be affordable to all, is based on the belief that the value of the telephone increases for everyone when everyone has access to a telephone. Decades before anyone imagined that the industry could be deregulated, government and service providers decided that universal service was its primary goal. Businesses paid premium rates to ensure that their customers could be connected to them. Discretionary services such as long distance were priced well over their costs.

The present pressure for service subsidies has not abated. Regulators established a universal service fund for the competitive market under the assumption that some customers would always require subsidies. Though local service, adjusted for inflation, has never been more affordable, and technology improvements promise to reduce customer costs further, subsidies show no sign of disappearing. New concerns that costly broadband services should be included in the basic service package could transfer more costs between customers.

Long-distance service rates have moved closer to their actual costs since the industry was deregulated, partly because of the reallocation of subsidies. Overcapacity in long-distance facilities pushed prices down further. Competitors entered the long-distance business by building some facilities and reselling in new markets until customer traffic justified investment in interexchange facilities. Long-distance usage and revenues grew quickly, and companies created an interexchange infrastructure with enormous capacity, which put downward pressure on prices.

Facilities-based local competition will not be likely at most of the present prices, unless fundamental changes to technology occur. Consumer local service is offered in most locations at prices well under its cost. The present structure for local resale is that the facilities owner sells to a competitor at a discount beyond its money-losing retail price, and the competitor resells the service to the customer. Many would-be competitors believe that the discount is insufficient to cover their marketing and administrative costs. There is little incentive for new participants to build facilities because the new facilities would increase their cost structure, not reduce it. Worse, since local service is often provided at a fixed price, creating additional usage will add costs but not revenues. This is particularly true in the consumer market. Internet usage, a major culprit, has increased the burden on the network but not the revenue.

Quickening the pace

One faded luxury for the monopoly provider was the ability to control not only what services were launched but when. In the interest of keeping local service rates low, technology introductions were deliberate and infrequent and had the tacit approval of providers and regulatory authorities. Competition would have ended this anyway, but increased bandwidth requirements demanded by customers hastened the service development cycle, due mainly to increases in online usage.

Within the telecommunications monopoly of the United States, technological innovation was important, but time to market was a lower priority. There were always regulatory issues to resolve, and their importance exceeded customer impatience for new services. It took 20 years to get cellular telephone service to the market, even after the technology was available. Depreciation schedules were designed to enable network providers to recover their investments, while keeping rates low. This lent itself to long equipment lives, sustainable only because no competitors were available. The monopoly mind-set does not lend itself to parallel product development cycles and replacing perfectly good equipment with slightly better equipment, but the technology mind-set does just that.

Customers used to the rapid obsolescence of computer technology will not tolerate preventable delays, and computer makers have risen to their challenge. Computers are obsolete long before they stop running, fostered by technology advances accelerated by competition. Gordon

Moore, the chairman of Intel, understood this when he predicted that computer chips would double in speed every 18 months indefinitely. Moore's law is still accurate, and customers expect it to persist, whether achieving the next increase in speed is smooth or troublesome for the developers. Technology customers expect improvement and lower prices to occur continuously. From semiconductors to timed telecommunications services, providers have responded to the challenge.

The deregulated telecommunications markets have proven that speed to market is not unique to the development of advanced technologies. They also insist that providers bring innovation in pricing, distribution, and promotion to market quickly. Customers are apparently passionate about simplicity in the rate structure, and they rewarded the long-distance providers that first met their desires with market power. Sprint was the first of the long-distance providers to offer a distance-insensitive rate per minute and gained brand recognition and market share from this innovation. AT&T fueled the already competitive wireless industry when it eliminated both roaming and long-distance charges from its digital wireless service. As with new services, the head start can only last until competitors design equivalent offerings. Still, both companies gained more market share with their bold pricing initiatives than any concern would have in second place.

The cast of players

Companies poised to provide a full range of telecommunications services are large and small, domestic and global, facilities-based, and resellers of facilities. Carriers entering new markets can choose to build or resell bundled facilities, lease and interconnect unbundled elements, acquire a carrier with an existing customer base, or create a marketing partnership. There is no formula yet for market readiness and no clear strategy for market success. They are all CLECs.

Still, the expected lineup includes several apparent categories of market participants. The deregulated market will comprise all telecommunications services, so the largest interexchange carriers (IXCs) and the incumbent local exchange carriers (ILECs) are obvious candidates. They have ready access to capital, millions of existing customers, and name recognition. They each will have to challenge an incumbent to win their existing customers' combined business; telecommunications customers very often subscribe to different local and long-distance providers. One provider's win of any customer is a loss to the other incumbent. IXCs have a significant advantage of experience in an intensely competitive market. Most have developed knowledge about how to convert prospects to customers and how to combat churn. The incumbent wireline provider already owns and manages the physical facilities connecting the customer to the network. This provides a cost advantage to the incumbent. The disadvantage of facilities ownership will emerge if much of their investment is stranded as market share declines to facilities-based competitors.

Another class of expected competitors is the facilities-based CLEC, which was once known as the competitive access provider (CAP). Several of the leading facilities-based local providers, including MFS, Brooks Fiber, and Teleport Communications Group (TCG), have been acquired by IXCs in preparation for local competition. Independent local providers include ICG; Intermedia Communications, Inc.; McLeodUSA; GST Telecommunications; and e.spire. Typically, these providers build high-capacity fiber rings around metropolitan areas, then connect customers to their own networks for a full range of telecommunications services. Most often but not always, they concentrate on high-margin markets such as businesses and multi-unit dwellings. This category is sometimes mistakenly assumed to comprise small, localized companies. IDC found that 12 companies generated nearly 70% of 1997 CLEC revenues, including MCI Local Service, a division of IXC MCI WorldCom [10].

Resellers constitute another category of local service provider, although it is misleading to allocate a separate category to them. Virtually every carrier will purchase services for resale somewhere at some time. Nevertheless, some providers are committed not to construct their own facilities. IXCs will resell incumbent facilities until the size or needs of their customer bases justify an investment in facilities. Local exchange companies will resell facilities in territories they do not currently serve. ISPs represent a group of strategic competitors in the telecommunications market and would undoubtedly serve local users primarily through resale should they enter the market. Local service providers look to the wholesale market as an opportunity to recover revenues from investments otherwise stranded. Researcher Frost & Sullivan estimates that the wholesale market for local services exceeded $14 billion in 1997.

Cable companies have made overtures to the local services market, but most of their residential customers are for high-speed Internet access in conjunction with an existing telephone line purchased from another provider. Speculation about coaxial cable as an alternate technology to the copper and fiber of telephone networks is an exercise fueled by active mergers and partnerships. If technology and investment barriers are overcome, cable companies will represent a different kind of facilities-based provider.

Leaders in the new marketplace

What will be unique about the winners in the deregulated telecommunications marketplace? Today's competitive markets provide some guidance, and *Marketing Telecommunications Services: New Approaches for a Changing Environment* is organized into the major elements of industry change. In general, it moves from the more general to the more specific and from the strategic to the tactical. Part I sets the stage. It focuses on where the industry has been, where it is now, and what generally happens to markets as they make the transition to competition. Part II focuses on marketing strategy. Each of its chapters covers a planning function that provides direction and market focus before tactical decisions are made. Part III describes alternative means to reach the customer. In a deregulated market, distribution strategies will vary among providers, even if each provider offers the same array of services. While long-distance and wireless providers are reaching out to new channels, this represents a significant change for many local providers that are accustomed to conducting identical distribution strategies in varied geographies. Part IV describes the challenges and solutions of differentiating telecommunications services in a crowded, near-commodity marketplace. Part V concerns the need for telecommunications providers to listen to their individual customers and develop programs to find them, satisfy them, and keep them. Not every telecommunications services provider will choose to compete in the area of customer focus, and some will be successful without leadership in this area. Those that take the route of customer focus will need to make a significant commitment to stand out in a field of high performers. Finally, Part VI describes the transformation required to ensure that the telecommunications provider is facing the market with all of its resources. All elements of the marketing-driven infrastructure, including the work itself, the cultural environment, and the technology support, need to realign to meet the needs of the marketplace. Market leadership requires focus and excellence in all marketing matters.

Each of the chapters covers one aspect of a changing marketplace. Market leaders understand the unique markets within which they operate. Today's local exchange service is a monopoly and will remain a near-monopoly in some geographical areas indefinitely after deregulation. In most places, though, market observers eagerly await the fall of regulatory restrictions and a competitive marketplace with new characteristics. At the dawn of local services deregulation, facilities-based interexchange service operates as an oligopoly, and the entrance of RBOCs will have an explosive impact on its characteristics. These issues are explored in Chapter 2.

The market leader of the future will have direction and focus. Monopolists are, de facto, the best at everything they do because nobody else is available for comparison. Successful competitive companies are never the best at everything; they choose their market positioning and become world-class in the core competencies required. This will require discipline on the part of prospective service providers. Most competitive telecommunications providers will need to scale back their broad mission statements and present world-class services to selected markets. The market leader will also recognize which tasks need full-time in-house attention and which can be more flexibly outsourced. Chapter 3 will cover these issues.

Market leaders will gather information rigorously about their changing environment and act on it decisively. Microsoft recognized the importance of the Internet and transformed its strategic direction, product development, and delivery systems in a matter of months. RBOCs pondered the "killer app" that would interest consumers, and when demand for second access lines arrived, their local networks, back office systems, and legendary customer service failed them. IXCs have been strikingly ordinary ISPs. The market leader will recognize the unusual and act in time. How to accomplish this is covered in Chapter 4.

Market leaders will target markets and develop service packages that uniquely appeal to individuals. They will utilize sophisticated network and information technology to serve millions of customers one at a time. Targeting market segments is a departure for the local exchange carriers (LECs) and somewhat of a departure for the largest IXCs. Outstanding market segmentation will be necessary if service providers want to compete successfully after the market is deregulated and the number of competitors increases. These concepts are discussed in Chapter 5.

Successful telecommunications providers will ensure that distribution channels advance their market strategies. AT&T's divestiture of Lucent Technologies was both necessary and well-timed. For AT&T, maintaining an in-house supplier became less of an advantage as the industry became more competitive, and competing with its customers was a long-term weakness. Decisions concerning which channels to serve are critical and will have a long-term impact on investment, customer segmentation, and overall market strategy. Strategies for optimizing channels are covered in Chapter 6.

Sales create customers. Some telecommunications providers will succeed because of their sales management, some in spite of it. The sales force of market leaders will be the right size and the proper mix, with cost-effective compensation structures. Misjudgments in planning and managing this function will result in lost sales and excessive expenses in sales, customer care, and churn. The direct sales function is discussed in Chapter 7.

Regulated telecommunications service providers are accustomed to having the in-house resources to run their businesses, and knowing—or controlling—their markets well enough to forecast their sales force size requirements. With new market entry, unknown competitive actions, and a dynamic customer base, service providers will find that the option of indirect sales gains a higher profile than it has in the past. Managing an indirect sales effort requires strategy and balance. These decisions are reviewed in Chapter 8.

The near-commodity nature of telecommunications services will compel providers to create strong brands for their companies and for individual services. While the largest telecommunications providers already enjoy company name recognition, RBOCs will be challenged to improve when they move into new geographical territories. Similarly, IXCs will need to improve their current branding strategies to unseat incumbents. Few of their branded consumer discount plans have distinct customer recognition. AT&T and MCI WorldCom have begun to develop product-based brands with their dial-around services, without an association to the parent company. Techniques to improve brand positioning are covered in Chapter 9.

The new market participants will decide whether to pursue a strategy of network technology leadership in a market in which all providers offer excellent quality. Sprint has led in promoting the quality of its network, and customers have repeatedly concurred, according to market research. Nevertheless, a 1998 Yankee Group survey ranked AT&T highest in six of eight quality-of-service categories, after losing the overall competition to Sprint for the preceding four years. AT&T's True Voice campaign leveraged its reputation as a high-quality provider, but its discount-based True Rewards program was more successful in acquiring customers.

Whether or not RBOCs choose to compete for technology leadership, they will still need to make significant investments in their networks. Those that decide to use their superior networks as a competitive differentiator will need to make commitments beyond historical levels to improving their infrastructures. Furthermore, unlike their regulated past, there is no guarantee that the networks will be utilized, that investments will be recovered, or that debt will be attainable at affordable rates. Chapter 10 addresses these issues.

It took the U.S. IXCs until 1995 to realize that customers would flock to per-minute, distance-insensitive domestic rates, a decade after AT&T's divestiture, and longer since competition began. Until then, rates varied by distance, by time of day, and by minute of call. According to Forrester Research, nearly two-thirds of consumers maintain that price is the most important criterion for selecting a service provider. Sprint pioneered low, simplified pricing, such as its "dime-lady" off-peak prices and its weekend flat rates. AT&T improved on Sprint's plan by offering flat rates with no time-of-day restrictions. Its rates were not the lowest, just the simplest. Pricing as a brand differentiator is covered in Chapter 11.

Surveys consistently demonstrate that customers also want simplicity by bundling services with a single provider. While the marketplace has never been able to test this assumption independently, most industry observers expect the service provider that offers the desired bundle of services at an unassailable price to lead the market. Chapter 12 reviews the research and the alternatives to determine the prospects for successful bundling of services.

Advertising is essential in all competitive markets, and as telecommunications become more intensely competitive, advertising expenditures have grown, and campaigns have become more sophisticated. Telecommunications providers need to develop advertising strategies that match their targeted market segments, their competitive position, and their management philosophies. Chapter 13 explores the many strategic decisions required of service providers in a competitive environment.

The experience of service providers in the competitive markets of long-distance and wireless has demonstrated that competing on price level, as opposed to other pricing characteristics, is an undesirable long-term market strategy. Prices continue to drop because of regulatory initiatives and technology, but these price decreases affect providers about equally. Most market leaders will choose instead to differentiate their services on other parameters. One issue is that price competition rents customers; it does not buy them, and customers courted by low prices churn as soon as a more attractive price is offered to them. To avoid the problems inherent in price wars, some telecommunications service providers will compete through superior customer care. Opportunities for leadership in customer care are explored in Chapter 14.

Most major carriers are investing in data mining technology, although few can quantify its benefits yet. Billing and customer databases maintained by telecommunications providers are massive and information-rich but cryptic. New information technologies enable telecommunications service providers to explore the marketing secrets hidden in the database. MCI's customer information technology is among the most sophisticated, not only in the telecommunications industry but also in any industry. The potential for customer data management technologies is described in Chapter 15.

Churn is a significant problem for service providers in any competitive market. While all carriers are equally afflicted by churn, AT&T will undoubtedly invest in its customer base, or it is at risk of losing more market share than it already has. Too many of its strengths are vulnerable, such as its critical mass, which will be equaled by RBOCs entering the market, and its loyal customer base, largely a result of customer inertia and a world-class reputation. Methods of customer retention are discussed in Chapter 16.

All telecommunications providers will need to eliminate any costs that do not further their market positions. Early indications of industry competition are that price will always be a primary buying criterion for the vast majority of customers. This near-commodity market will not tolerate a high cost profile.

Most fundamental business processes, especially those related to marketing, will require transformation. Local telecommunications providers entrenched in a monopoly mind-set will undergo significant cultural change, as will the interexchange providers whose market until recently has been limited in scope and limited in competition. The need for change and its impact on the industry is discussed in Chapter 17.

MCI WorldCom earned its reputation as a trailblazer in competitive services in the earliest days of long-distance competition. Its market positioning strategy has been to leverage its commanding market presence with a superior application of information technology. Telecommunications providers still primarily utilize technology to conduct routine functions or to maintain competitive parity. There are few examples of telecommunications providers using their networks as a marketing tool. As technology companies, telecommunications providers lag most industries in every information technology category except cost. Chapter 18 describes the benefits available through the strategic use of technology and the need for providers to make a commitment to their application.

References

[1] Henderson, Khali, "Smaller IXCs Gain Market Share, FCC Study Shows," *Phone+ Online,* October 5, 1998, www.phoneplusmag.com.

[2] Mason, Charles, "PCS Carriers Leverage Price Advantage," *America's Network,* Vol. 102, No. 18, pp. 19–20.

[3] Parker, Tammy, "PCS Pileup," *Telephony Supplement: PCS Edge,* October 26, 1998, pp. 4–7.

[4] Taylor, Jennifer, "Elusive Customer Loyalty," *Telephony,* Vol. 234, No. 17, p. 72.

[5] Koppman, Steve, "Long-Distance Services," *tele.com,* Vol. 2, No. 13, pp. 35–36, 38.

[6] Hirschman, Carolyn, "Fearsome Creatures," *Telephony,* Vol. 234, No. 11, pp. 18–24.

[7] Hill, G. Christian, "Telecommunications (A Special Report): Bypassing the Bells—Consultant's Call," *Wall Street Journal,* Eastern Edition, September 21, 1998, p. R27.

[8] Verger, Jose, "Competition With a Capital 'C'," *Telephony,* Vol. 235, No. 16, pp. 62–66.

[9] Krapf, Eric, "Is Competition on Your Horizon?" *Business Communications Review,* Vol. 28, No. 9, pp. 44–48.

[10] Perrin, Sterling, "The CLEC Market: Prospects, Problems, and Opportunities," *Telecommunications,* Americas Edition, Vol. 32, No. 9, pp. 26–30.

2

Competitive Markets

The transition to competitive markets

Any analysis of the telecommunications market needs to highlight the transition from a regulated monopoly market to deregulation and full competition. The effect of deregulation in the long term is unknown, as no telecommunications market in the world is yet fully deregulated. It is clear, though, from the experience of markets that have begun deregulation, that companies moving from regulation to competition will transform themselves in every aspect of their business practices, especially in their staffs. The ideal employee of a monopoly, in any business function, is probably not the ideal employee of a fully competitive company, and vice versa. Competition is already causing some industry consolidation in the United States and Europe that will have an impact on the culture and identity of these merged entities. Some of the most significant differences will occur in marketing because the entire structure of the telecommunications market is changing.

Monopolists, including telecommunications providers, always need to understand the nature of markets, but they can generally confine their analysis to customer behavior and not the actions of competitors. Moreover, monopolists have as much time as they need, within reason, to react to external market forces. Managers of regulated monopolies do not need to react quickly to the twists and turns of their industry, or respond to competitor actions. A successful monopoly does not change much.

Competitive markets, however, change constantly. Fortunately, the behavior of competitive markets has been an object of research for decades, and much of what is known about them has already been demonstrated in the emerging telecommunications industry. The telecommunications industry can benefit from the marketplace knowledge of similar, competitive industries, such as the computer services industry and other industries that have undergone deregulation, including airlines.

Impact of deregulation

Deregulation affects every business process undertaken by a telecommunications provider, but no function is affected more than marketing. In the past, the success of telecommunications providers depended on skills and attributes other than marketing. Among such attributes were companies' access to capital, their management of relationships with regulators and consumer groups, and their provision of unilaterally consistent services and quality. Marketing was primarily an external relations support function, involving the regulatory process, the company image, and groups representing customers. Products and pricing were developed in closed committee sessions, and the services offered to the market embodied many objectives. Customer demand was among the objectives, but it was not necessarily the most important.

The marketing function supplies the window on customer needs. Marketing will be central to the survival or success of telecommunications providers in the future. Accordingly, marketing professionals will need to work closely with operations groups to ensure that required service levels are met. They will work with engineers and planners to ensure that their sales objectives correspond with capacity plans. In addition, they will need to consult with finance professionals to ensure that prices, costs, and sales levels result in company profitability.

Companies undergoing deregulation need to modify what they do, the number of employees required to accomplish these tasks, and the skills and assets required of employees. Marketing organizations will need to grow in size. In parallel, other functions will reduce the importance of certain activities, and the employees formerly involved in those functions will then be able to be absorbed into the growth of marketing. The following sections describe the most visible changes brought about by deregulation and their impact on the marketing function.

Price competition

Regulated markets set prices that are nonnegotiable and not subject to competitive pressures. Conversely, competitive markets revolve around price. The significance of price competition will affect virtually every activity of the telecommunications provider. Senior marketers will need to understand the mechanics of pricing, which means that they should have a fundamental comprehension of both microeconomics and finance. They will be able to use corporate resources to conduct detailed analyses, but a basic understanding will help marketing professionals to target the most profitable segments and create the most effective pricing strategies.

The introduction of price competition will require all market participants to monitor, and often emulate, the pricing tactics of competitors. Consequently, companies will need to develop information and operations support systems to respond quickly to competitors' price changes.

When companies compete fiercely on price, services in high demand cannot subsidize higher cost services without regulatory intervention. Services priced too high will lose market share; services priced below cost will gain customers but lose profits on every sale.

Reduction in the cost structure

The most important ingredient of pricing a service is the cost of providing the service. A service's price has to be within the customer's willingness to pay, but the price must exceed costs for a service to be viable as an ongoing offering. Price competition means that telecommunications providers will need to eliminate all unnecessary costs from their cost profiles.

Monopolies tend to generate excellent, consistent services. To accomplish this quality and reliability, and to prevent customer service complaints, monopolists tend to err on the side of making additional investment and incurring higher costs than companies in competitive markets.

Some cost reductions will occur in functions that are no longer required in a competitive market. In the transition from regulation to deregulation, some organizational divisions will shrink. This will occur if their workload, such as meeting regulatory reporting requirements, diminishes. Other divisions will need to grow and to build their own supporting information systems. The marketing function is most likely to grow.

Other less visible costs influence the monopoly cost structure. Monopolies that provide higher-than-market salaries or better-than-market benefits packages will be competing against firms with much lower costs. Monopolies, because of their low risk, command relatively low interest rates on debt and equity funding. Risks are higher in competitive markets, so interest rates will rise, which will increase the cost profile. In other words, a regulated company whose cost profile today is equivalent to its unregulated peers will be at a disadvantage when financial markets account for additional risk in an unregulated market. Furthermore, regulated companies often enjoy higher debt ratios—that is, debt as a percentage of overall funding—than the general market. Debt is less expensive than other funding alternatives. When the debt ratio decreases to an average market level, overall costs will rise.

Competitive marketers constantly strive to provide acceptable quality at the lowest possible cost. In markets with ardent price competition, competitors shave costs to the minimum to retain market share. Marketers in the competitive telecommunications arena will need to balance the established need for quality service with the added need for low prices.

Direct focus on customers

In a monopoly market, providers recognize customer needs from two directions: the actual customer and the customer's representative in the form of regulatory oversight and consumer groups. Because the customer has no alternative sellers, there is not very much to learn from either their buying behavior or the hypothetical perceptions of their behavior that are obtained through market research. The customer of a regulated monopoly does not have a context in which to evaluate purchases. The customer could resent the amount of a monthly local service bill but be unable to provide an accurate assessment of his or her own willingness to pay. Instead of a customer focus, the regulated monopolist has a customer image. Monopolists can aim products toward the average needs of customers without fear that another provider will come closer to the target.

In a competitive market, customers have a choice of providers, so knowing the individual needs of customers is critical. Customers are more informed about buying choices and the prices of each alternative. Product introductions are driven less by technology and more by the customer's stated desires. One benefit that competitive providers have is that the market itself provides direct confirmation of new strategies and nearly immediate notification of market failures. Customers have more confidence in the value of their own purchases when services are offered by several providers.

Shortened product cycles

Competitive markets sustain shorter product development cycles and shorter product life cycles overall. Product development is hastened because providers are racing against their competitors. Being first to market with a new product can often command a premium price and procure economies of scale before competitors are prepared to launch their versions of the product. New techniques for product development have arisen to meet the needs of industry, especially in technology-driven industries.

The first-to-market company can actually hurt its entry by rushing the product to the marketplace. AT&T launched its WorldNet Internet service in the hope of attracting customers with its blue-chip brand name. The timing was good because the Internet was just beginning to attract consumers in large numbers. AT&T offered free software and free service to new users, the typical incentive offered by ISPs. AT&T was not prepared for the response from prospects. First, the company ran out of software at its New Jersey headquarters. Worse, the new subscribers that managed to get the software installed met with busy signals, slow response times, and a customer service group that was unprepared to solve problems. Later, WorldNet sustained a five-hour outage in its e-mail service. In its zeal to benefit from being in the market first, AT&T may have caused its own problems. The credibility of the AT&T name, which drew the customers in the first place, can exacerbate the customer annoyance when the service does not meet expectations.

Product-based profitability

To meet social objectives, monopolies intentionally price products without a clear relationship to their costs. Competitors cannot afford to continue this practice without the assistance of regulators. Long-distance service was always priced well above its costs in the United States. When AT&T owned most of the local telephone companies, AT&T overcharged long-distance callers and sent most of the profit back to the local companies, which undercharged their subscribers for local service. Regulators did not want to eliminate this subsidy when other long-distance providers were admitted into the market, so they simply applied the same subsidy to the new entrants.

While this price shifting can work for the short term, eventually regulators discover gray areas that can undermine their plans. Internet telephony is emerging as a low-cost threat to traditional long-distance service. In an effort to encourage the growth of the Internet, most American regulators have avoided applying taxes and subsidies to this form of communication. Eventually, this will result in a lopsided price structure in which one service has artificial costs applied and the other does not. At that point, regulators will choose between imposing subsidies on the new service or phasing them out of the traditional one.

Industry consolidation

Historically, industries undergoing deregulation have undergone a consolidation phase. The consolidation occurs when an industry cannot support the sheer number of participants it holds. In some industries, including telecommunications services, economies of scale require the participation of huge providers to achieve profitability.

Smaller players in regulated industries can be sustained by revenue-sharing arrangements overseen by regulators. They need to find strong partners when the subsidies decline or disappear. In the airline industry, some carriers that steadily served unprofitable routes with regulatory support did not survive the shakeout of deregulation.

For telecommunications, consolidation in both the long-distance and local markets shows no sign of slowing. Of the original seven regional companies divested from AT&T, five remain. Two large mergers are in process, which will absorb Ameritech and independent telecommunications provider GTE. The requisite size for a telecommunications provider is not yet known. If critical mass turns out to be a necessity, the companies that chose not to merge will be at a disadvantage in the markets served by their larger colleagues.

MCI and WorldCom initiated the consolidation in the long-distance industry. Strategic alliances create the mass of consolidation without the commitment of a legal union. AT&T and British Telecom (BT) launched a major marketing partnership in 1998. Its purpose is to provide global communications services through their existing networks and new technologies.

Other forces changing the market

The transition from monopoly to competition is not the only significant factor affecting the telecommunications market. Other drivers include the advancement of technology, the world economies, globalization, and changing demographics.

The rapid *advancement of technology* has driven the charge toward competition. At the same time, technology's rapid pace is partly a result of competition. When monopolists controlled the entire telecommunications industry, they also controlled the pace at which communications technology was developed and introduced to the marketplace. Monopoly-funded research at Bell Laboratories and other facilities provided the research foundation for the information age and the subsequent development of products and services in competitive markets. Competitive markets do not enjoy the luxury that monopolists possess to manage the introduction of products to the market.

The convergence of computer technology with telecommunications has increased the demand for telecommunications services. Personal computer usage has pushed the development of communications technology in the form of networking, modems, and new applications for wireless communications.

The relative strength of *world economies* and globalization can work together to foster or hinder the pace of competition and its nature. Half of the world does not have telephone service at all, and this population represents a new customer base—if pricing and logistics can be resolved. When economies are strong, global commerce thrives and becomes fiercely competitive. When large economies founder, all other economies are affected. The growth of the telecommunications industry in any country will continue to depend on the fortunes of other international economies.

Globalization affects the nature of competition as well. Barriers to international commerce are descending. Whereas each country might only support one or two domestic telecommunications giants, a world market enables each giant to participate in a huge market. Marketing alliances between national telecommunications companies and international mergers and investments are evidence that a world telecommunications market is emerging.

Changing demographics are affecting telecommunications markets in many ways. Family and business structures continue to evolve, and technology enables further changes in the way people live and work. Telecommuting, wireless technologies, and paging technologies free customers to live and work in ways that were impossible without these conveniences. Small companies can act like big companies, and big companies can become gigantic and at the same time more personal. These changes, enabled by technology, will alter the nature of telecommunications competition.

The structure of markets

Economists have identified four different market characteristics: *perfect competition*, *monopolistic competition*, *oligopoly*, and *monopoly*. These market types are of great interest to telecommunications marketers because the industry is undergoing a transition from monopoly to each of the other types. Different service lines and segments will undoubtedly mature into different structures because of the inherent differences in the products and the customer base.

Several factors determine a market's structure, including the number of sellers, the size of each seller, and the way prices are controlled by sellers. Markets generally display certain characteristics, and competitors in different markets generally perform as expected in each market's form of competition. Knowing what to expect from competitors and buyers helps telecommunications providers adjust their marketing strategies to be successful in newly competitive markets.

Markets with *perfect competition* have many sellers providing homogeneous goods, and well-informed customers who have access to many sellers. Any single firm cannot control the price of the product because customers are aware of the lowest price offered, and it is easy for customers to go to a nearby competitor.

Markets characterized by perfect competition demonstrate high *elasticity of demand*, which simply means that when the price is lower, customers will buy more, and when the price goes up, customers will buy less. Sellers that dare to drop prices within perfect competition will benefit in two ways if the new price is still profitable. First, their own customers will buy more of their product, and second, customers of their competitors will switch because the price is lower and the product is similar. Of course, if the seller drops the price to an unprofitable level, the seller will lose.

Sellers always have the option of reducing their level of output if their profits do not meet their objectives. While the actions of one seller do not affect the market as a whole, when all sellers reduce their output, prices will rise. When sellers increase capacity, prices will fall.

The long-distance resale market is an example of perfect competition. There are thousands of providers. No single reseller has the power to control prices, and the product is a near commodity. Overcapacity in transmission facilities has encouraged resellers to produce more output, that is, sign more customers by offering low prices. The demand for interexchange services has also grown as the prices have dropped, demonstrating elasticity of demand.

In *monopolistic competition*, there are many sellers, but the products are perceived by customers to be relatively heterogeneous. The unavailability of one seller's product cannot be met by another seller. For this reason, competitors in monopolistic competition have a limited ability to change their own prices, but they cannot affect the overall price in the market.

How does a company make its product or service appear to be different in the eyes of the customer? In service businesses, any feature of the product can act as a differentiating factor. Customers who purchase their Internet service from their local telephone company have chosen not to buy the exact same service, resold at a small discount, from a new industry entrant. The customer perceives a difference because of the service provider. (Methods for differentiating products are discussed in Chapter 9.)

Prepaid card-based long-distance services are provided in monopolistic competition. Providers (most often resellers) compete fiercely for the best rates in general, but they can collect higher prices when they offer the cards in creative locations, specialize in a certain customer group, or design and distribute collectible cards. Some providers of international services obtain favorable rates to a specific country, then market the cards in neighborhoods that attract immigrants from those countries. Their domestic rates are within an average range, but the international rates differentiate the card. No single provider can change the overall rates of the industry, but a prepaid card seller can increase a company's sales when it targets a specific population, or increase its rates through a superior distribution channel.

Oligopoly markets are characterized by a few large competitors that control most of the output in an industry. Price changes are often initiated by one of the market participants, and the others either choose to follow suit or not. Once all the competitors have raised their prices, customers have little choice but to buy at the new, higher price. Demand is *inelastic*, that is, price changes across the industry do not result in a significant change in the volume sold. A single provider's unanswered increase or decrease will affect that provider's share.

Pricing in an oligopoly becomes a strategic game. Before the first participant raises its price, it analyzes what the competitive response will be from the other participants. If the other sellers do not raise their own prices, demand for the first seller's product will decrease, and the price increase will reduce profits, not increase them. The seller then loses. If, on the other hand, the other sellers match the price increase, everyone's profit increases, and customers are forced to pay more for the same product. Often, one participant in an oligopoly tests the market by raising its price. If others do not follow suit, the provider reduces it to the former level.

Cartels are oligopolies, but they are not legal for most commerce. Cartels intentionally agree that products will be sold at a certain price. Sometimes, cartel members agree on the price, then one partner reneges on the agreement and sells at a lower price. When the product is a commodity, the errant seller increases its sales significantly at the expense of the other cartel members.

"Follow-the-leader" pricing has been in evidence in the facilities-based long-distance industry, where the market is controlled by only four or five providers. Collusion among firms, however, is illegal, and most governments are vigilant at preventing or punishing it.

Wireless markets, due to licensing requirements, are oligopolies. Agents for the facilities providers can package services in creative ways, but they cannot change the fundamentals of industry pricing.

Telecommunications services were provided in a *monopoly* or near-monopoly structure for their first century. Worldwide, governments are moving the industry toward alternative structures to monopoly.

Not all monopolies are regulated. Even in competitive markets, monopolies exist, and many are legal. A monopoly exists when a single company dominates and controls the prices in an industry. The monopolist does not need to control 100% of the market.

For example, a software company could develop a PC- or handheld-based package that telecommunications providers could use to test a specific type of circuit on the network. If no other firm makes a similar package, or if the software developer does not actively use its market power to thwart competitors, the software provider enjoys a monopoly, but a legal one. Microsoft, a somewhat large software developer, provoked an investigation concerning monopoly power; however, the company can demonstrate that it does not control 100% of any market it serves.

The legality of monopolistic practices is serious business, and marketers need to be aware of the risks they take. Marketers that are concerned about crossing any of the limits would be wise to engage an attorney who specializes in antitrust law.

In a fully deregulated world, legally sanctioned monopolies will probably persist in geographical areas that are simply uneconomical to serve by more than one provider. Pricing in these situations will undoubtedly fall under some form of regulatory scrutiny for a long time.

The forces driving competition

A highly respected work describing competitive markets was written by Harvard professor Michael E. Porter. The book, *Competitive Strategy,* provides a framework upon which to analyze any industry, its players, and the forces behind change and success. Porter's premise is that competition in an industry is the result of basic forces outside of the control of the competitors that participate in its activities. These forces are depicted in Figure 2.1.

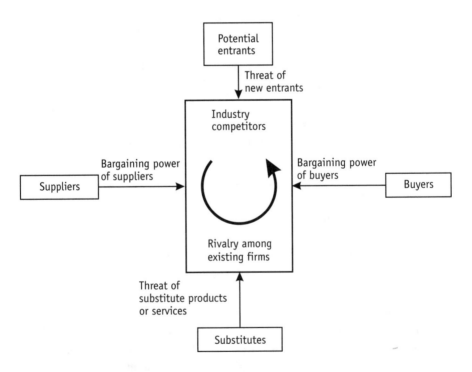

Figure 2.1 Porter's model of industry competition.

As depicted in Porter's diagram, the *intensity of rivalry* among competitors is central to understanding the competitive nature of an industry. An industry shows substantial rivalry when there is significant price competition, when advertising expenditures exceed those of similar industries, or when there are high *barriers to exit* for the existing participants. Exit barriers are present when unprofitable participants do not leave the industry. A telecommunications provider with a losing division in, for example, paging services might need to stay in that market to offer service bundles with its profitable wireless service. This inability to exit the industry raises the number of competitors and increases its competitiveness. While a high level of intensity is difficult for the sellers, customers generally benefit from lower prices.

The U.S. long-distance telecommunications segment has demonstrated intense rivalry, and it meets several of Porter's expected characteristics: Many sellers entered the market as soon as they were permitted; the products are relatively equal; customers are able to switch providers easily; and capacity is high. This assessment is substantiated by the fact that the amount of advertising by the largest long-distance providers matches the annual expenditures of makers of other highly competitive products, such as soft drinks. The ubiquitous telemarketing call to request that the customer switch long-distance carriers has become a cultural icon in the United States and a source of mockery by television comedies.

Rivalry in the wireless market, while competitive, has always been less intense than competition in the long-distance market. This was especially true in the early development of the wireless industry. The characteristics that create an intense marketplace—many sellers, large capacity, and unencumbered customers—were not present: Each geographical area had a maximum of two competitors; limited capacity was available; and customers were bound contractually for a relatively long period.

The *threat of entry* will affect competitor behavior because new competitors bring resources, added capacity, and a desire to earn market share. When entry is easy, competitors flow in until the market is not as profitable as alternative businesses. The facilities-based telecommunications market carries high *barriers to entry*, but resale does not. Most facilities-based telecommunications businesses have considerable economies of scale. Scale economies create profitability for the largest providers, but they result in high costs until a certain level of volume is met.

To become a facilities-based provider, the new market entrant needs access to a great deal of capital. This serves as a barrier to entry, which keeps new entrants out, and a barrier to exit that holds incumbents in. A large investment discourages those providers that have already built infrastructure capacity from abandoning the investment and leaving the market.

Frequently, there are regulatory barriers to overcome. Wireless providers require licenses, which are expensive and have a long lead time. A 14-point checklist has served as a barrier to entry into long-distance markets for RBOCs. The items on the checklist were designed to facilitate local resale and interconnection before enabling the incumbent LECs to compete for long-distance service.

Patents serve as a barrier to entry for many products and are likely to assume more importance when the telecommunications services market is more competitive. Stakeholders with an interest in retaining the present structure can present barriers for new entrants. Telecommunications providers and consumer groups often take initiatives in the press or in the courts to block new entrants from their markets or to prevent the companies that serve them from diversifying.

Access to distribution channels can act as a barrier to entry or, viewed positively, can act as a facilitator to entry when distribution channels are widely available. IXCs that want to sell local services, and local service providers that want to sell long-distance services, have ready access to their existing customer base. Having distribution channels in place will facilitate their entry. The potential market entrant that does not already have a customer base needs to develop distribution channels from scratch, and this serves as a barrier to entering the market.

The *threat of substitute products* is another market force. Products can be considered substitutes if they have similar performance characteristics, if they have similar occasions for use, and if they are sold in the same geographic market. All three characteristics must be met for a product to serve as a substitute for another. Satellite telephones are not a substitute for pay phones, even though they both provide telecommunications service and may be sold in the same town. Satellite, because of its complexity and cost, is not idly used by the person who wants to call for a ride home. The occasion for use (maritime or emergency services, for example) is totally different from that of the pay telephone service. This explains why the price commanded for satellite telecommunications is very different from a similar-length, similar-distance pay telephone call.

Substitutes for the products offered by an industry segment tend to create additional competition simply through their availability. When PCS was introduced, it was not an exactly equivalent product to cellular service, but it closely resembled its function. For very mobile users, cellular service coverage was better when PCS was launched, although PCS had some service advantages as well. In any case, aggressive selling by PCS providers resulted in more intense competition in the wireless industry.

The growth of wireless services, advanced text-based paging technology, and even electronic mail can act as substitutes for public telecommunications. As wireless services become increasingly affordable, they will make an impact on certain segments of the pay phone industry over time.

For most long-distance users, technology has presented an array of options that serve as substitutes. Facsimile is often used as an alternative to personal calling. Electronic mail is emerging as a viable substitute for both calling and fax—and often without any incremental cost to the user.

In the consumer long-distance segment, the discretionary use of long-distance service acts as a substitute, in the sense that customers have the option to avoid making calls when the call's value is not obvious. Business users generally do not have this choice.

Strategically, Internet-based telephony represents a long-term substitute for traditional circuit-switched communications services. Telecommunications providers need to be aware of emerging substitutes for their core products and be willing to select one direction or both as dictated by corporate strategy. Sprint recognized the importance of PCS as a wireless substitute for cellular but was prevented by regulators from holding PCS and cellular licenses in certain geographical areas. The company spun off 360° Communications, its cellular division, and concentrated its forward-looking efforts on PCS.

In a monopoly, there is little *buyer bargaining power*, which is why regulatory authorities are needed. In competitive industries, buyers have varying levels of power to influence pricing and product development.

Industries characterized by high-volume buyers demonstrate the most bargaining power for the buyer. An interexchange wholesaler that sells capacity to only three resellers is at a disadvantage compared to the reseller that sells to millions of consumers. The loss of any one of these customers can have a significant impact on profits.

Buyer bargaining power is not confined to the largest customers. When the cost of a product is a large share of overall costs, buyers are very careful shoppers, and assert bargaining power. A personal computer buyer will assert more bargaining power than the same individual purchasing a cable for the same computer. Buyers who can either buy the product or make the product in-house can use that capability to hold a seller accountable and negotiate favorable prices, reducing the seller's profitability.

When selecting market segments to serve, the strategic decision needs to include an assessment of the bargaining power of buyers. The buyer who exercises less power is often more loyal and less price-sensitive than the buyer with power is.

Supplier bargaining power can intensify the rivalry in an industry. For facilities-based telecommunications providers, suppliers have enormous power. Facilities-based providers can purchase their most costly and strategically important network equipment from only a few suppliers, domestic and international. Often, these purchases involve long-term contracts and maintenance agreements, so the telecommunications provider is committed for a long time once the purchase decision has been made.

For resellers of interexchange service, many transmission facilities providers are available and highly competitive. Time commitments for purchases are much shorter than for facilities-based providers. Suppliers to long-distance resellers thus have less power.

The reseller of local service, on the other hand, has fewer options, at least for the near future. In most markets, the reseller needs to purchase service from a single incumbent provider, at a price that is often established by tariff and is nonnegotiable. When multiple facilities providers are available in local markets, supplier power will decline, the bargaining power of the buyer will increase because of the availability of choice, and this will cause supplier prices, and eventually local service prices, to fall.

SELF-ASSESSMENT—COMPETITIVE MARKETS

Answers to the following questions will help marketers analyze the deregulated market in which their companies operate.

- What is the structure of your market?

- What can you do to take advantage of the characteristics of the market, such as barriers to entry or exit, or the power of buyers and suppliers?

- What can you do to minimize the disadvantages inherent in the market structure?

- Is the market moving from one structure to another? Is it moving from monopoly to deregulation or from monopolistic competition to consolidation into oligopoly? What is your role in the transformed market structure?

- What are the barriers to entry in the new markets you intend to serve? Are your plans realistic?

- What is the intensity of rivalry in new markets? If it is higher than in your current markets, can the company meet the challenges of the new markets?

- Is the cost structure of your company competitive? What changes are being made to ensure that costs are kept low? Are these changes in keeping with the overall market strategy?

Part II

Assessing the Market

3

Marketing Planning

The need for a marketing plan

"The last thing IBM needs right now is a vision." So stated Lou Gerstner upon taking the helm of the struggling computer giant in 1993. While observers at the time debated the wisdom of that position, no further vision statements did emerge from IBM, and with the company's impressive recovery, the dividends have been quite visible.

The meaning of Gerstner's statement, quoted and analyzed by many industry observers, was that IBM needed fundamental skills in operations and marketing to succeed. At the time, IBM's costs were higher than those of its competitors, and it needed to regain the customer focus for which it had been famous. Visions are alluring but perhaps insufficient.

The same problem faces telecommunications service providers as they enter the world of deregulation. The planning and strategic skills developed over a century of service are crucial but incomplete. Marketing in a highly competitive market has not been among the necessary tools in the monopoly tool kit. Developing a marketing plan will help to set priorities and objectives, assign responsibilities, and organize resources. A marketing plan is not a vision; it is a map.

Raymond Smith, former chairman and CEO of Bell Atlantic, kept a small sign in his office that said "Execution IS Strategy." The less philosophical and more pragmatic the marketing plan is, the more achievable it will be. A marketing plan needs to be updated with actual results and appropriate corrections to reflect changing markets. To the extent that frontline employees and customers can help to shape the plan, the plan will benefit accordingly.

A marketing plan is a contract between the marketing organization and the parent organization. The marketing function requests a certain level of funding and support and, in turn, promises to deliver a certain level of production in categories such as unit sales, revenues, or market share.

The marketing plan is most useful if it is not produced in an ivory tower. Contributions from frontline sales representatives, large and small customers, and marketing managers all improve the accuracy and usability of the plan.

When completed, the marketing plan (edited for highly sensitive material) should be made available to all marketing employees and potentially to all company employees, each of whom undoubtedly contributes to the marketing effort and would benefit from this knowledge.

Goals and objectives

The marketing goals and objectives represent a commitment to producing certain measurable results during the planning period. Marketing plans contain a definition of the company's business. This definition is driven from the business plan, which undoubtedly contains a company's mission statement and a statement of the business scope. Mission statements occasionally suffer from excessive breadth, and the statements of scope often err on the side of comprehensiveness.

According to press releases and other public materials, most of the telecommunications giants do not plan to limit the scope of their present businesses. At the same time, they are vocal in their desires to enter new businesses. Local wireline companies intend to offer long-distance service, expand geographically, and offer content such as video to their customers. Long-distance providers want to provide facilities-based local service, expand internationally, and offer more broadband services to customers. While it is possible that some of the current providers will accomplish these goals, it is less likely that all of them will do so successfully without reducing the scope of their business in some way.

Marketing plans need to meet the objectives they set; it is appropriate to limit the scope of the business for the planning period to define short-term goals. Market planners who make revenue and sales commitments need to forecast realistically within the planning horizon.

Marketing goals, then, should have the following characteristics:

- Marketing goals need to be a practical subset of corporate goals. They should address all of the achievable scope of the business and describe appropriately scaled long-term plans to enter the businesses that are strategically important but not yet viable.

- Marketing goals do not need to include numerical results, as do the marketing objectives. Nonetheless, they should be worded clearly enough so there is no doubt as to whether they have been met at the end of the planning period.

- Marketing goals should be supported by the rest of the marketing plan.

The next section of the plan includes the marketing objectives. Like goals, objectives represent the desired market position to be achieved during the planning period. Objectives differ from goals in that they are measurable and more specific than goals. While goals present the overview of the provider's desired position in the marketplace, objectives spell out the various metrics that will determine if that position is achieved. Objectives that are presented in the marketing plan will be the focus of managers and their employees; if a desired objective is omitted, it will probably be left unattended.

Objectives can include revenue targets, both aggregate and for individual products; revenue levels from certain market segments or geographical areas; and unit sales in the aggregate or in certain segments.

They can include strategically important measures such as new accounts, customer retention, major account management, customer service, customer satisfaction, or leadership. Companies that recognize that a product is nearing the end of its life cycle can include objectives to change the sales mix and realign its proportions toward newer products. Companies can choose to measure sales to the installed base as compared to new customers.

Objectives can specify internal measurements such as new distribution channels or the reduction in channels, sales productivity, information systems, or employee turnover in essential customer functions. In general, each objective will include the desired numerical indicator, how the indicator is measured, and when possible, the current measurement of the indicator. It is appropriate to set objectives that are less favorable than today's measurement. A local exchange company that is expecting a significant increase in competition should set a market share objective lower than the present level. This objective should represent an achievable, but not leisurely, level of customer retention.

Secondary objectives are generally included without strict numerical metrics. These objectives support broad corporate goals such as technology leadership, service quality, or a customer focus. On the other hand, when these goals represent primary corporate objectives, they too require distinct objective measures.

Situation analysis

The situation analysis provides the context in which the plan is to be reviewed. This section describes the internal and external environment surrounding the planner's decisions and expectations.

The overview of a situation analysis should enumerate the various relevant macro factors that will affect the company's performance in the marketplace. Macroeconomic indicators such as inflation, global markets, consumer debt, and the overall state of the economy will affect customer purchases and thus affect market size. Porter's five forces model, described in Chapter 2, provides a framework for evaluating the forces affecting the industry.

Demographic changes can diminish the success of certain products in favor of other products. Demographic changes can create the need for new products. Without strong market planning, demographic changes can diminish a telecommunications provider's own success in favor of its competitors.

Technology affects the marketing plans of all industries, but it is a most significant macro factor in the telecommunications services industry. This section of the marketing plan only considers the broadest implications of technology on the marketing plan. Other sections can cover how specific technologies will be marketed, or company-proprietary technologies that strengthen the company's market position. Still, this description provides the strategic backdrop for the marketing plan.

The introduction of a new technology or the improvement of an existing technology might eliminate the market for an existing technology. Integrated services digital network (ISDN) struggled to find its market for years. By the time the market could have embraced the technology for fast Internet access, 56-Kbps modems and faster technologies such as digital subscriber line (DSL) were on the market or around the corner. While certain applications are well-suited for ISDN and the technology has not disappeared, the large target market that was anticipated for the service never materialized.

Political and legal influences, including the regulatory environment, should be described. A provider that is unrestricted by regulations that affect its competitors should assess the anticipated impact when the restrictions on other companies are lifted. Regulatory restrictions can affect the timing of an anticipated market entry. Few U.S. telecommunications providers predicted the delay between the signing of the Telecommunications Act and the emergence of a fully competitive market.

If the company is dependent on a specific supplier, or a specific type of supplier, any potential disruption to the supplier or its industry needs consideration. Companies that buy bandwidth from IXCs should consider the state of overall capacity during the market planning process. Overcapacity will keep downward pressure on prices, but it can also encourage a favorite supplier to exit the business. Insufficient capacity can result in higher prices during the planning period, or service outages.

Among the more important components of a marketing plan are estimates of market size and the company's share within the market. Both size and share should be stated in three ways: today's, growth in the last five years (or a desired number of years), and anticipated future growth.

Market size is generally stated as annual sales and can be qualified to a single country or continent. Market size can be stated in terms of production or consumption. As critical as this knowledge is, estimates of market size vary, and a marketing plan should present the most accurate and qualified total, as the market size can be the basis for some sales forecasts shown as a percentage. An estimate of size or growth that is too ambitious could result in an unrealistic sales forecast. On the other hand, an underestimate of market size could strain both marketing resources and the company's infrastructure.

In the mid-1990s, local telecommunications providers did not anticipate the growth in second residential lines resulting from the popularity of the Internet. In many areas, installations were delayed beyond customer expectations. Customers waited for days or weeks for second-line installations. Furthermore, some local wireline companies recognized the sales opportunity for second lines and promoted them actively without coordinating their sales plans with the operations groups responsible for installations.

The inability to estimate growth can also have an impact on operations and customer satisfaction without affecting sales or increasing revenues. In one Bell Atlantic study, Internet usage was responsible for a fourfold increase in "all circuits busy" announcements and problems reported grew from zero to 25 per month [1]. It is the responsibility of the marketing plan to anticipate these trends. They can then provide early warning to the other corporate functions that are in a position to accommodate them.

While the rapid growth of the Internet exceeded most analysts' expectations, those companies that incorporated accurate estimates in their marketing plans were better positioned to provide excellent customer service and enjoy a more rapid revenue stream.

Market share is calculated by taking a company's revenues (or sometimes units sold) as a percentage of the total market size. A $1 billion company has a 5% share in a market of $20 billion. Not surprisingly, market share is difficult to calculate. The denominator, market size, is not an absolute figure. The market share of competitors is very difficult to compute, especially when the competitor is a division of a larger company. In addition, companies are often reluctant to provide their own revenues or unit sales to public venues if they are not required to do so.

Calculating one's own sales for the numerator is quite simple, but sales need to be formulated on the same basis as the calculation for the market size denominator. If the company includes, for example, overseas sales in its own financial statements, then the denominator must include world market size, or overseas sales must be subtracted out. Other less apparent distinctions are also necessary and will depend on a company's own record-keeping practices.

An analysis of the installed base (or present subscribers, for service providers) can yield useful information about changes in market share. The service provider whose share of new customers is larger than today's market share is growing in share. A drop in share is predictable when companies cannot achieve a rate of new customers to match their share of the installed base.

AT&T lost about 50% of its market share in the first decade after long-distance deregulation and its divestiture of the operating companies. While this drop could be devastating for companies in most industries, AT&T's profitability did not suffer. Advances in technology, the unprecedented increase in long-distance calling, and management cost controls all helped AT&T to withstand the reduction in share.

Growth, while a useful metric, requires normalization. If the company changes its business scope during the period of review, the historical numbers will require recalculation. Maintaining a historical growth rate is always a challenge, but many market planners increase the rate of growth as well as the base revenue forecast. A common technique evident in business plans or marketing plans is the "hockey stick" growth trend, where "growth until today" is greatly exceeded by "growth after today" without further explanation. See Figure 3.1.

While some marketing plans do support an increase in the rate of growth for a product or service, "hockey-stick" trends are justifiably viewed with skepticism from senior management.

The marketing plan often includes a section describing the company's current market positioning. This serves several purposes. First, it provides a baseline for the proposed market positioning objectives and helps

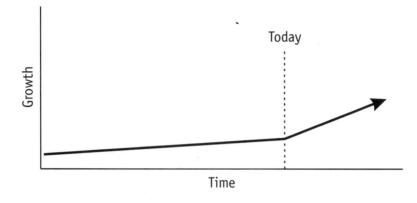

Figure 3.1   "Hockey-stick" growth.

management to evaluate whether the objectives are realistic. Second, the positioning analysis describes competitors' market positions, their perceived strategies, and spending indicators, if they are available. The spending indicators include information such as how much money competitors invest in infrastructure, information systems, advertising, and research and development. The competitive spending estimate provides at least a starting point for spending decisions in the same categories, proportional to company size.

A customer analysis describes the target market or markets for the service. The marketing plan should contain the proposed market segmentation strategy and the analysis that supports it. Customer needs change, and the plan should reflect the known differences since the most recent prior analysis. The customer analysis can contain emerging trends about customers, either through changes in demographics or lifestyle changes that affect telecommunications spending, and new groups of potential customers that might have been overlooked in the past. This is also an opportunity to list customer needs that are presently unmet by any provider.

Similarly, a product portfolio analysis describes each product or service, its profitability, and its future. Information about each product can include its sales level, the proportion of this product to overall sales, its profitability, historical and anticipated growth, cost trends, marketing investment, and trends in distribution.

The marketing plan can assign each product to a phase in its life cycle to alert management that a substitute is needed. Products at the end of their cycle often warrant a change in their marketing. New investment is not warranted, and resources assigned to its support should be assigned cautiously. Gaps in the product line are as important as the products themselves. Opportunities to sell new products should be described in detail, including a consideration of their impact on sales of existing products.

Channel analysis evaluates distribution channels for each market segment to determine what actions to take to maximize sales. Each channel is evaluated for its effectiveness and its cost. Channels can be ranked to determine which is producing the most sales in the aggregate. New channels can be tested, or existing channels phased out. Resources should be added to productive distribution channels.

A similar analysis can be performed vis-à-vis the sales organization. If the amount of data maintained is sufficient for analysis, the attributes of the most successful sales representatives can provide useful information for modifying the composition of the sales staff. By product or by segment, certain characteristics, such as sales training or the type of compensation plan, are often determined to be more productive than other alternatives. Over time, the productivity of the sales effort can be optimized.

Identifying strengths and weaknesses

Strengths, weaknesses, opportunities, and threats (SWOT) analysis, a popular method for assessing the environment and a company's position within it, simply evaluates these four elements. Strengths are positive attributes that are internal to the organization. Weaknesses are negative attributes that are internal. These attributes can include access to capital, marketing skills and experience, distribution channels, operational efficiency, leading-edge technology, cost control, or service quality. A telecommunications provider might be strong relative to its competitors in the area of marketing but relatively weak in access to vital distribution channels. Strengths and weaknesses are meaningful only in comparison to similar characteristics held by competitors.

Opportunities and threats refer to the attributes of the outside marketplace or factors outside of the organization's direct control. Opportunities are positive and external; threats are negative and external. Opportunities include new markets, regulatory relaxation, or competitors exiting the marketplace. Threats include the planned entry of strategic competitors, anticipated economic downturns, or the turnover of large customer contracts.

Ideally, opportunities can emerge from threats. Internet-protocol (IP) telephony is potentially the most significant threat and the most important opportunity for most telecommunications providers for the next decade. The market planning decisions now being made in boardrooms will determine whether IP telephony is a threat, an opportunity, or a fad.

For a SWOT analysis to be most effective, it should limit its scope to the most important success factors in its own marketplace. An ISP that serves a local market may not have the software distribution budget available to America Online, but that factor does not matter unless it hinders the company's ability to compete with America Online in the chosen market.

Similarly, the analysis should exclude those factors that, while important, do not distinguish the provider from its competitors. RBOCs consider their access to capital to constitute a strong point, and it is, but only as compared to less mature competitors. Most of their major competitors are not small entrepreneurs, but one another. The provider would need to reconsider that attribute in an analysis that lists its competitors to be other RBOCs and the world's leading interexchange companies. For a set of competitors with an equal access to capital, this is a weakness, not a strength. Strengths indicate superiority, not adequacy.

The usefulness of a SWOT analysis increases when it is updated on a routine basis. This provides not only a snapshot of the company's position at a given time but an indication of its forward or backward momentum. A continuous review of the SWOT factors also provides a window on the rising and waning importance of the various criteria for success. At some point in every market, certain success factors rise in importance and others become less important.

Some planners produce a political, economic, social, and technological (PEST) analysis, which evaluates the external environment in terms of the political, economic, social, and technological factors. This analysis is especially valuable in the present telecommunications industry because deregulation and changing technologies will challenge most of the assumptions currently used in the market planning process.

Early warning of newly critical factors can help a company be prepared for marketplace changes. Microsoft recognized, before many other technology companies did, the growing importance of the Internet. It retooled all of its software products to be more Internet-aware. Whether Microsoft's sheer size was partly responsible for the rapid growth of the Web, or whether the company simply demonstrated excellent market planning, is now unimportant. Its skill at forecasting, coupled with its ability to turn operations around in a matter of months, greatly increased its marketplace success.

Marketing strategy

Once a comprehensive analysis is complete, the marketing plan can recommend actions that meet the needs of the marketplace, within its constraints, taking into consideration the strengths the provider brings to the market. The market strategy defines the actions to be taken by the provider during the planning period to achieve the defined objectives.

Much of the substance of the marketing plan is a review of knowledge that has been acquired on a continuous basis. In some ways, the written plan is a snapshot of the day the plan is released. After the plan is released, the marketing strategy and its resulting tactics are developed on an ongoing basis, as opportunities arise or as modifications to existing plans are deemed necessary.

The product and brand development strategy defines activities necessary to create services that are viewed as unique and valuable by customers. The activities required to accomplish this strategy often involve functions other than marketing. Those responsible for its completion are consulted during the plan's development for their commitment to the task.

Pricing strategies need to address an assortment of variables. Customer willingness-to-pay, product development costs, operational and maintenance costs, and a decision concerning flat rates versus usage-based rates all need to reconcile in the pricing strategy. In some cases, pricing requires approval by regulatory authorities. Both the likelihood of approval and the regulatory time lag need to be accommodated by the plan.

Channel strategies are spelled out, especially if new channels are involved or if channels are to be phased out. Channels should be associated with the market segments they serve, including a justification for each channel, based upon superior customer access, low cost per revenue dollar, or effectiveness.

The marketing plan will describe promotional and advertising plans, although it cannot serve as a comprehensive document. These plans, which need to be coordinated with the overall marketing strategy, would be fully described in a separate document. In any case, corporate advertising campaigns, and, at the next level, segment or product strategies, are always planned at the highest level of interfunctional cooperation feasible. Advertising executives should be represented in every phase of the marketing planning process to present a coordinated campaign and assist in identifying market needs.

The marketing plan should also address the changes to infrastructure that are required to accomplish the objectives. These changes include additional staff, training programs, information system support, and other marketing tools. The plan may include a proposal for a market test to evaluate a new product, enter a new territory, test a promotion, or try a new channel. Each year, the marketing plan can earmark a certain amount of funding to conduct real-time research to pursue new markets.

The market strategy cannot overlook the likelihood of competitive response to the provider's actions resulting from the plan. Revenue forecasts must account for a slowing of demand when competitors respond with their own price reductions, promotions, or new products based on innovations in the marketing plan.

Sales forecasting

Sales forecasting is among the most difficult tasks of the market planner. In the dynamic telecommunications industry, forecasting is complicated by the timing of deregulation initiatives, new technologies, consolidation of the industry, and the direct impact of the macro factors described by the plan's situation analysis. Still, forecasts are required and provide the basis for marketing activity in the planning period.

The most accurate forecasts use a variety of sources, conduct a top-down and bottom-up forecast, and provide at least three forecasts from most conservative to most aggressive. Top-down forecasting begins with the most general data, such as projected inflation rates, historical growth patterns, and general economic conditions, and extends the trend for company overall sales before subdividing by line-of-business or other categories.

Bottom-up forecasts assemble the estimates of sales professionals, beginning with the sales representatives, and then moving on to divisional, regional, and national levels. Eventually, the forecasts are combined into a single estimate. Depending on the structure of quotas and compensation plans, this technique can produce inaccurate results. Knowing that the forecast will drive sales quotas, sales professionals might tend to underreport their expected sales. Sales professionals with less rigid compensation plans could over-report their opportunities. Territories not clearly assigned can produce double counting of the same customers. After several rounds of forecasting, the market planner will be able to compensate for the biases inherent in this technique.

Customer loss is sometimes overlooked in sales forecasting. Forecasters can focus more on the sources of growth and assume that the customer base remains stable. Large customers are lost through their mergers with others. The gains made from acquiring the business of a customer's merger partner often do not compensate for losses of merged customers to competitors. Mergers are frequently justified for their economies of scale, including their use of telecommunications services. Small companies and individuals are lost to marketing campaigns of competitors. The telecommunications services business, the largest contracts aside, is not characterized by significant switching costs. Churn in the industry is at record levels. Research conducted on a year-to-year basis will be useful in estimating customer loss for forecasting purposes.

The technology adoption life cycle, shown in Figure 3.2, is a bell-shaped curve that depicts a pattern of purchasing technology in the marketplace. Typically, when technology is first introduced, only the innovators acquire it. As its benefits become known and its ease-of-use improves, its growth rate increases, and the early adopters become cus-tomers. The growth rate continues to increase until the market begins to become saturated with the early majority and late majority. Eventually, the laggards join the others.

Figure 3.2 does not depict the product life cycle, and the technol-ogy's growth probably does not wane when the laggards join. By the time the late majority and laggards are buyers, the innovators and the early adopters are probably upgrading or doubling their usage of the technol-ogy. Instead, Figure 3.2 is useful as a tool for sales forecasters. Accurate estimates of a technology's position on the curve and the length of the cycle will help the service provider calculate accurate forecasts of sales, choose effective distribution channels, and develop appropriate enhance-ments to the services.

The sales forecast should clearly identify any planning assumptions that underlie the data. Management can evaluate the results in the context of the assumptions supporting the data.

Augmenting internal capabilities

During the market planning process, the company sometimes discovers that its internal capabilities are not sufficient to meet the defined goals and

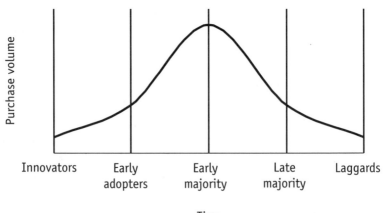

Figure 3.2   The technology adoption life cycle.

objectives. The SWOT analysis may conclude that the company's direct sales skills will not achieve the sales objectives for a given target market. One option is to eliminate the sales objective for that market and from the aggregate objective. A more favorable option is to obtain the needed resources to achieve the objective. Acquisitions and mergers, alliances, and outsourcing represent several ways to accomplish this.

Acquisition of a smaller company or merger with a similar sized partner is a popular way for a company to grow quickly. Strategic mergers and acquisitions can also provide new skills and marketing assets to companies. Industry observers widely believe that a competitive telecommunications industry cannot support the dozen or more existing giants, so more mergers are inevitable. Telecommunications providers apparently believe this as well. Most of the regional Bell companies, GTE, AT&T, BT, MCI, and WorldCom have all completed or considered mergers with another large telecommunications provider. By the end of 1997, more than 470 mergers involving U.S. telecommunications companies had taken place since the signing of the Telecommunications Act of 1996 [2]. Industry consolidation is virtually a certain outcome of deregulation.

Acquisitions are permanent commitments, and success is far from a guarantee. AT&T acquired NCR and divested it, almost certainly after a significant loss. Some mergers fail before they are completed, such as the attempted coupling between Bell Atlantic and Tele-Communications, Inc. (TCI). Regulatory approval poses a substantial barrier. For companies that want some of the advantages of mergers with almost none of the headaches, marketing alliances are a better option.

Marketing alliances between companies attempt to expand territories, cross-sell services, or consolidate marketing efforts without the burden of corporate restructuring. AT&T and BT announced a marketing alliance in mid-1998. Most of the major international providers, such as Concert, Global One, and UniWorld, are involved in alliances. Marketing alliances can serve as a precursor to mergers, but not always, as BT learned when it tried to acquire its partner MCI.

Research demonstrating that 60% of new residents to an area call the electric utility before the telephone company led BellSouth and other carriers to experiment with alliances with the local utility [3].

One disadvantage of alliances is indeed the lack of commitment from both partners, which makes separation easier but prevents the sharing of private corporate data. Like all partnerships, a sense of imbalance can occur when one partner believes that its contribution to the union is greater than its reward.

Outsourcing has become popular as companies decide that they cannot excel at every required activity. Companies tend to outsource functions that they do not believe they require as a core competency. While most companies outsource functions until they are large enough to bring them in-house, telecommunications providers are accustomed to meeting most or all of their corporate needs with in-house staff. In a highly competitive market, telecommunications providers will need to be world-class in virtually every function to succeed. This goal will probably be achieved only when some functions are outsourced.

While marketing is indeed a mission-critical function, certain components of the function can be outsourced successfully when controlled by management. Telecommunications providers, as part of their SWOT analysis, are certain to identify weaknesses in some tasks that are mission-critical. Telecommunications providers have added to their marketing arsenal by outsourcing the telemarketing channel and some customer service. Successful outsourcing requires a clear definition of the provider's requirements, unambiguous measurement criteria, and continuous management attention. In short, success in outsourcing is the same as it is in any other supervisory function.

Contingency planning

As sophisticated as a marketing plan might be, it is only a set of expectations. All plans need to include allowances for contingencies. Contingency planning is not as simple as adding a 5% fluctuation to each revenue estimate. It also involves a careful review of the factors identified in the situation analysis and speculation about probabilities and possibilities. What if the company's planned acquisition doesn't go through? What if the biggest competitor becomes much bigger through acquisition? What if the plan's estimates on the timing of new technologies are wrong? What if a competitor delivers a product that cannot be matched in the proposed portfolio? Each of these scenarios can generate discussion about preventive measures, reactions, and monitoring techniques, often without the need to change the substance of the plan.

SELF-ASSESSMENT—MARKETING PLANNING

Here are some questions to assist telecommunications marketers in the development of a marketing plan.

- Does your company have a marketing plan?

- Is there a formal planning process, with due dates and senior management approval?

- Does your company use its marketing plan?

- Are company managers measured by the objectives set in the marketing plan?

- Does the plan's development utilize all of the resources that would improve it, such as interviews with front-line sales representatives or customers?

- Is an adequate level of contingency planning being performed?

- Does anyone review the entire plan after the planning period with the intention of improving planning techniques?

References

[1] Snyder, Beth, "Pileup on the Public Network," *Telephony,* Vol. 231, No. 9, pp. 28.

[2] Lawyer, Gail, "Give and Take ... and Take," *tele.com*, Vol. 3, No. 2, pp. 64–68.

[3] Docters, Robert G., "Corporate Alliances: Do You Really Know Why You Are Getting Married?," *Telephony Supplement: In Focus*, Vol. 234, No. 15.

4

Competitive Intelligence

Types of competitive intelligence

Competitive intelligence, competitive analysis, and business intelligence are all terms that describe a company's attempt to learn about the markets it serves and its own positioning within them. Competitive intelligence is used to predict what competitors will do before they do it and react accordingly to the knowledge. While competitive intelligence primarily targets competitors, the same investigative techniques can provide valuable insight into the plans of suppliers or major customers. Unanticipated changes affecting important suppliers or customers can have more impact on a business than the actions of individual competitors, so they are worth watching carefully. If a company's largest customer is acquired by a company served by its competitor, it will either have a larger customer or a lost customer after the acquisition takes place. The earlier a company knows about the merger plans, the better its chances are of earning the combined company's business.

Interest in competitive intelligence has grown in the last few years. When companies learn that their peers are engaged in management programs such as competitive intelligence, they often establish their own programs, if only to maintain parity. In the telecommunications market, any increase in the intensity of competition will undoubtedly increase the prominence of competitive intelligence programs. Ironically, because some companies do not publicize the extent of their competitive intelligence efforts, other companies may inaugurate programs believing that such programs exist in competitors' organizations.

Strategic intelligence concerns itself with understanding competitors' long-range plans, the future of the competitive market, and the potential entrants to the marketplace. Strategic intelligence is by nature more generalized than other forms of intelligence but no less urgent. This form of competitive intelligence can provide insight into potential mergers and acquisitions that can drastically alter the competitive landscape. For example, did BT assume that its marketing alliance with and partial ownership of MCI would forestall any other merger offers? Did the company predict the wave of interest in MCI by WorldCom (its eventual partner) and GTE?

Tactical or *operational intelligence* has a shorter time perspective than strategic intelligence. Tactical intelligence attempts to learn about competitors' short-term pricing plans, capacity changes, and service introductions. Companies need to be actively seeking information to glean tactical intelligence in time to react.

Counterintelligence is the other side of competitive intelligence. Because counterintelligence is the process by which companies defend their internal information, employees are the barriers between this information and the intelligence efforts of competitors. Another counterintelligence necessity is the erection of computer firewalls and controls against hackers.

While much of competitive intelligence gathering benefits the marketing function, the data can come from many sources within a company. A firm's counterparts to the division that holds the targeted company's desired data are the most efficient intelligence-gatherers; its operations staff reads the trade journals that contain the information about competitor operations; and its information systems staff can learn about its competitors' use of strategic technology.

Information that is gathered in a competitive intelligence program also needs interpretation by personnel in other organizational functions. Financial statements are valuable sources of information but only to people who know how to read them. Often the most revealing insights are buried in the financial notes. If competitive research uncovers that a competitor is building a facility of a certain size, only an operations supervisor could know what the physical capacity of the new building could produce in sales. Thus, interdepartmental cooperation in the analysis of the data is advantageous.

It is difficult to summarize the results of a competitive intelligence program within a single page or report. To acquire the full benefit of such a complex program, several additional steps are required. The competitive intelligence analyst, with support from managers of other departments, needs to synthesize the data and make assumptions as to the future behavior of competitors, suppliers, or major customers. It is then management's job to propose and implement an action plan. Until the entire process is completed, the company cannot expect to benefit from its competitive intelligence investment. Implementation contains the payback. The process is depicted in Figure 4.1.

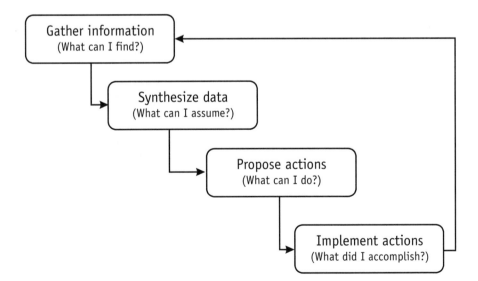

Figure 4.1 The competitive intelligence process.

The scope of the competitive intelligence process suggests that some companies that attest to maintaining a competitive intelligence program fall short in the later phases. Some gather the information and make an effort to synthesize it but then do not develop action plans. Others make recommendations and fail to implement them. Many companies admit to having no formal program at all.

In 1997, The Futures Group conducted its third annual survey of competitive intelligence efforts in American business. All respondents stated that they gathered intelligence at some level. Only 60 percent were found to have an organized business intelligence system. Most respondents represented companies with more than $1 billion in annual revenues. Eighteen percent of companies with revenues of more than $10 billion did not have an organized program for business intelligence.

The telecommunications industry has been well-represented in the top 10 companies making good use of competitive intelligence. Motorola was awarded a position as first or second by its peers in 1995, 1996, and 1997. AT&T was in the top 10 in both 1995 and 1996.

To promote interactions between its business units, BellSouth has hosted a formal competitive intelligence program, sponsoring monthly meetings, a database, and paper communications [1]. The director of strategic intelligence at US West Media Group was recruited from the Central Intelligence Agency and army intelligence [2].

As expected, the most visible competitive intelligence programs are found in the largest and most powerful companies. Still, smaller companies with competitive intelligence programs will benefit as well. In fact, smaller companies are more likely to have a single powerful supplier or a few important customers, and are more likely to be significantly affected by the actions of one company. This makes them more likely to benefit substantially from an investment in a competitive intelligence program. Unlike many other management tools, competitive intelligence can be established on a relatively small budget, requiring only one or two full-time professionals. Since much of the research is now conducted on the Internet, a large investment in research tools is not required.

The important first step is to do something. Woody Allen once said, "Eighty percent of success is just showing up." An organized competitive intelligence program can help a company reduce costs, manage its pricing strategies, and create services demanded by the marketplace. It has been estimated that most of the competitive information a company needs is already known within its own organization. Locating and sorting this base of information and responding with appropriate actions can add substantially to the bottom line. Apparently, "just showing up" can be a competitive advantage.

Competitive intelligence requirements

Today, all companies have nearly the same access to competitive information. More than any other time in history, industry information is published with wide, even international, distribution by more sources than have ever existed in the past. Much of the information about competitors is free, or the cost is negligible. The biggest hurdles are making the commitment to managing competitive intelligence, drawing appropriate conclusions from the information that is gathered, and acting on the conclusions.

Information has a short shelf life. By the time most market information is well-known, it has little value other than being a historical record. A competitive intelligence program needs to have the mechanism to use and act on information the company finds. This requires a full-time competitive intelligence effort and a commitment by management to listen to the findings, make decisions, and act quickly.

Monitoring competitive developments consistently will yield more insight than last-minute research assignments. Much of the most useful information is transitory, in a press release or a brief news item, and could be lost if not captured when it first occurs. A constant window on the behavior of industry competitors also provides balance and context to conclusions and analysis.

Competitive intelligence requires a willingness to resist optimism when faced with unpleasant findings. There are many ways to interpret a piece of information, and most valuable findings do require interpretation. The interpretation that recommends business as usual is often tempting, but can be inaccurate.

In 1996, the International Benchmarking Clearinghouse (of the American Productivity and Quality Center) conducted a study to understand what activities comprise the most successful competitive intelligence programs. It learned that the best intelligence programs had the following practices in common:

- With senior management involvement, they focused on areas that are critical to operational success.
- They formalized the program and responsible individuals, then managed the knowledge that they collected.
- They institutionalized the program into the daily activities of managers.
- They changed the corporate culture to achieve organizational goals and objectives.
- They used their competitive intelligence successes and failures to improve the program.

A systematic competitive intelligence program can offer substantial returns on a relatively small investment. Competitive intelligence provides an early warning signal about competing products, perhaps in time to mitigate the potential damage by developing a new generation of products in response. These programs can help companies avoid costly investments in untested markets by revealing that more entrenched providers are eyeing the same territory.

A search for intelligence data can uncover a competitor's plans to exit a market or sell a division, raise or lower prices, or introduce a new technology.

Competitive intelligence can cover virtually any aspect of the company's operations, from marketing and product intelligence to production capacity or distribution plans. Copies of corporate organization charts are sometimes available for sale. These charts can reveal both the corporate strategies of competitors and their organizational weaknesses.

When embarking on a competitive intelligence program, it is important to validate data whenever possible through a second set of sources. Some data are irrefutable, such as revenues published in an annual report. Other data, published by sources whose goal is to persuade, can be accurate but incomplete. For example, trade associations provide very accurate data most of the time. Occasionally, they need to lobby on behalf of their members. The supporting data they provide are probably accurate, but these associations cannot be expected to provide all of the data that support an opposing point of view. For that data, refer to the opposing trade association, which has its own viewpoint.

For the analyst, a computer equipped with a modem and basic business software is sufficient to begin a competitive intelligence program. A more ambitious plan would use the company's internal network to support the transfer of competitive information between departments. An employee who notices something new or unusual about a competitor could prepare an electronic mail message to the competitive intelligence analyst. A more sophisticated system could include a questionnaire or form for employees to complete and send. Information systems support should make it as convenient as possible for employees to pass along information.

Sources of information

For companies under $1 billion in revenues, *business and trade publications* are the best sources for obtaining timely information. To learn about the direction of competitors' business, such companies can also recruit internal information systems staff to read trade journals and report to management about the sophistication of competitors' information systems. To determine competitive advertising budgets or programs, companies should subscribe to advertising and public relations magazines, extending their sights beyond their own markets and reading the publications targeted to their suppliers and customers. Publications and marketing materials written by a company's suppliers may boast about large sales to the company's nearest competitors. Similarly, trade journals targeted toward a firm's customers' industry will provide information about customer satisfaction with the firm and with its peers.

General publications can be useful, as well. Publications targeted to investors and the business community often profile companies or industries. Local newspapers in competitor locations can provide insight. The help-wanted ads can often provide an ongoing glimpse of competitors' fortunes and insight into their growth plans. Advertising and promotions provide information about products offered locally.

Government documents are well-researched and often free. Governments collect data about industries and require publicly held companies to file financial reports. In the United States, the Department of Commerce collects industry information, such as its size and growth statistics, and summarizes the data into reports.

In the telecommunications industry, useful statistics are gathered by regulatory authorities. The Federal Communications Commission (FCC) in the United States and the Office of Telecommunications (OFTEL) in the United Kingdom gather and summarize information provided by telecommunications service providers in their jurisdictions. The information provided contains important intelligence that is difficult to obtain elsewhere, such as market share and market size.

Government documents are a useful source of financial information for privately held companies. Private companies, unlike publicly traded companies, do not post their financial information with the Securities and Exchange Commission (SEC). Private companies provide otherwise-unavailable financial data in court hearings, government proposals, or applications for government assistance. This information then becomes available to the public.

For significant business transactions, companies leave a paper trail. Permits, loans, and participation in local hearings can provide information about competitors' growth plans. These documents are often available in government offices.

The *Internet* has become one of the best sources of business information in history, and much of the information is free. The main advantage of the Internet is that keyword searches can review thousands or millions of pages in seconds and retrieve the pages that meet the request. This introduces another success factor—the ability to create searches that retrieve desired documents and eliminate unwanted ones.

Of course, when information is this readily available, all competitors have the same access. Still, the company that concentrates its effort on research, and couples the information gathered with intelligent analysis, will gain competitive advantage.

The quality and accuracy of pages on the Internet vary greatly, and the Internet researcher needs to be cautious and critical. Pages, even prepared by companies and trade groups, are often designed to persuade. Information provided on a page can be true but incomplete. Internet research should be validated by multiple sources and ideally by impartial ones.

Trade and business publications often host sites on the Web that mirror their printed publications. Many also provide a search engine, making it possible to retrieve archived material using keyword access. General business publications sometimes charge a nominal fee for the retrieval of archived articles.

Business news wires post press releases from companies, trade associations, and government. One advantage of monitoring the press releases issued by individual companies or their representatives is their timeliness. The disadvantage is that they need to be interpreted through a filter of the company's objectives. A company can post a release saying, "Our company just invented the best product," or "Our company has met all requirements to approve this merger." Trade associations, like their member companies, might exaggerate the truth or omit relevant unfavorable data to make a point. This information is better than having none, but proper skepticism is warranted.

Other Web-based services provide more information and analysis for a monthly fee, a fee per search, or a combination. Often these services provide access to commercial databases, opening windows to investment reports, trademark information, and credit information. Internet clipping services that comb the Internet daily and tag new pages for clients whenever they find keywords requested by the client are also available.

Market research firms can be engaged to assist in answering specific questions about a company's competitive position. These customized studies can be expensive, but if they provide the answer to strategic questions, they are well worth the investment. Market research firms that specialize in the telecommunications industry often create detailed reports and sell them to interested customers for thousands of dollars. While these reports are not inexpensive, they provide nearly the same amount of value as a customized study for a fraction of the price. Market research companies also publicize summaries of their research on their own Web sites and in business news wires.

A cost-effective competitive intelligence program usually takes the route from the least costly information sources to the most expensive. The first path through market research would probably go through the publicly available documents on the Web. Next, the analyst can interview the market research firms that look most promising and request additional materials. Sometimes the firms will provide printed summaries of their research reports, which is enough in the areas where detailed data are not necessary. For areas in which further detail would be valuable, the analyst can purchase the research report. After the analyst has a complete understanding of what information is available at no cost or for a low cost, custom research could still be warranted.

For telecommunications providers selling services to consumers, a company's own *employees* can help to uncover competitive marketing strategies. A long-distance service provider can ask employees to send a survey form, via internal electronic mail, each time the employee is solicited by a competitor. The survey could contain questions about the competitor offering the service, the type or price of services offered, or the frequency of the solicitations.

New employees hired from competitors or from suppliers can provide a new view of the market participants. While some information is off-limits, and it is critical that employees understand this, new hires can lend insight into a competitor's corporate culture, management style, and morale.

The ethics of competitive intelligence

It is important to state strongly that competitive intelligence is not espionage. Professionals who perform competitive intelligence adhere strongly to guidelines that prohibit the violation of any laws. The ethics of competitive intelligence gathering should be part of a company's corporate compliance program. Any employee or agent engaged in competitive intelligence activities should be familiar with the company's code of conduct.

Most of the competitive information that is truly valuable is not protected as a trade secret. A competitive intelligence program does not need to break laws or enter gray areas to be successful. Theft of trade secrets is illegal and clearly not part of an acceptable competitive intelligence program. As to specific practices, the law varies by jurisdiction about what is legal and what is not.

Some types of data and some legal data gathering techniques are still inadvisable. Sometimes the reason is the appearance of illegality. For example, a competitor can discuss its marketing philosophy to a group of listeners in an industry seminar. In a private discussion after the formal session, a company's representative and the competitor could continue the discussion, which can appear to violate antitrust laws or easily become a violation. In such a case, it would make sense to give up the added information and the potential appearance of impropriety. The potential cost of defending borderline competitive intelligence practices, even if the responsible parties are cleared of wrongdoing, is not worth the risk. The incident would also not reflect well on the company if its occurrence were revealed to the public. Competitive intelligence programs should be designed with the assistance of senior management and legal counsel. Whether professional analysts or not, employees who are encouraged to find and pass along competitive information need to understand their legal and ethical boundaries. Employees conducting competitive intelligence would benefit from training on the law of trade secrets and antitrust, as part of a company's overall compliance program.

Questions arise as to whether to identify oneself or one's motives when conducting competitive intelligence. Notwithstanding any legal exposure, an overzealous analyst could resort to behavior that the sponsoring company does not sanction. The Society for Competitive Intelligence Professionals (SCIP) has published guidelines for the profession. SCIP recommends that analysts accurately disclose all relevant information, including their own identity and the organization they represent, prior to all interviews.

In addition to this guideline and the others on the SCIP list, companies can use rules of thumb as a corporate policy to assist analysts in making ethical decisions. The analyst should ask, "Can my behavior cause harm to our company?" or "Would I like to see this behavior on the first page of the newspaper?"

Competitive company analysis

A wealth of company information is available for publicly traded corporations. Learning about smaller companies takes more work. The smaller companies with a significant market presence will appear in trade publications and the local press in their headquarters location, even if the company is privately held. They might sponsor a Web page or appear in documents posted on the Web. Companies whose market visibility is so small that they do not advertise on the Web, and are not covered in even the local press, are probably not worth monitoring.

Corporate publications often provide a fruitful start for competitive intelligence efforts. Companies provide a wealth of information on their Web sites, including annual reports, organizational charts, product descriptions, promotions, and texts of executive speeches. Companies are often torn between the need to inform investors and potential customers and the fear of revealing information to competitors. The advantage of Web research is its convenience and ability to search. Even if the information is also available in stores or in the printed annual report, the Web is a great time saver.

The U.S. government's *Edgar* database, searchable on the Web, holds all of the financial filings of public companies traded in the United States. This database, maintained by the SEC, is free of charge. Every quarter, companies file financial reports, and often the reports include information about company activities or company strategies.

A competitive intelligence analyst can take the reported data and calculate comparison ratios for industry leaders by evaluating certain categories of expenditures as a percentage of revenues. Numerous ratios can be developed using the number of employees and costs or revenues in different categories as either the numerator or the denominator. For these ratios to be sound, the raw data for the various companies must be analogous. A local exchange company serving primarily rural areas will have a higher cost structure than a company targeting urban businesses. Large companies serving a broad base of customers probably apportion their costs differently from small niche providers. Data from disparate companies must be normalized for analysis to be valid.

Securities firms have teams of analysts who follow the telecommunications industry and some of its segments. These analysts provide *industry and company reports* to their customers, primarily to recommend certain stocks or industries. These reports are well researched, and they uncover exactly the same information that a competitive intelligence analyst needs. They publish reports about individual companies and comprehensive reports about industries. Each report contains information such as estimates of industry growth or the market share of companies currently in the industry. Because investors are interested in long-term prospects, these reports cover strategic issues, including the anticipated effect of technology, regulatory issues, or emerging competitors.

Investment analysts do not follow some niche telecommunications markets when they do not directly affect the value of any specific stock. For example, the U.S. prepaid phone card market was not analyzed on its own when it first emerged. The cards were most often offered by private entrepreneurs or company divisions that did not amount to a sizable percentage of the parent company's portfolio. Most providers were local and long-distance telephone companies, pay phone providers, or money transfer service providers. Once several companies that specialized in phone cards went public, they became a "pure play" and began to get attention on their own.

Large telecommunications providers with many divisions generally report revenues—but not income—for each division. A company entering new markets sustains a loss in that division for several years and is not eager, or required, to report the loss. Investment reports do not provide hard data, but they can offer some insight into how close a given division is to being profitable.

There are dozens of telecommunications *industry conferences* every month. Most are open to anyone interested in attending (and interested in paying the registration fee). Typically, companies send representatives to these conferences in search of customer leads, and those representatives make presentations about their products, accomplishments, and internal business processes.

It is true that trade secrets are not generally included in the presentations, but many companies are candid about their strategies. Some companies provide case studies about their customers and provide data about the size of product lines or geographical divisions. Much of this information is not reported on financial statements. Occasionally, companies will use conferences and trade shows as vehicles to announce new products and their pricing. Suppliers that make presentations at the show can provide insight into the competitive plans of their customers. Sometimes the show's management will provide taped or printed transcripts of sessions for a fee.

Competitive product analysis

There are many ways to evaluate competitors' products and services, many of which are not routinely used by companies seeking information. One such method involves first identifying which products are truly competitors and then targeting them for analysis.

One way to see the competition firsthand is to *buy competitive products and services* and have employees use and report on them. For tangible products, companies have learned to conduct *reverse engineering*. They physically take the product apart and study its construction and capabilities. In the service business, the process of reverse engineering is a metaphorical one. It starts by buying and using the competitive offering.

The evaluation can include an analysis (and comparison) of the brand name, its features and packaging, distribution channels, pricing, promotion, customer care, and advertising. This will enable a company to learn the following:

- What is different about the competitor's service?

- What aspects are better? Can our service be modified to meet this challenge?

- Do customers seem to use competitors' products more often or differently from our own? Why?

- Are the customers in a different market segment?

- Is some feature of the competitive product making it easier or more convenient to use?

- Does the competitor call to ensure the service is satisfactory?

Analysts can learn about competitive offerings and industry positioning by *reading published reviews*. When markets are contested, trade magazines, especially those that target users, often conduct annual tests comparing products. Internet magazines and periodicals routinely compare the top Internet service providers. Surveys compare pricing, connection rates, and response times and even gauge customer opinions of content. For the most part, these analyses are impartial and are conducted with statistical rigor.

Some services serve so small a market that industry publications do not publish reviews about them. In these situations, market research firms can *conduct competitive studies*. Suppose a company provides frame relay services only to educational institutions. It is unlikely that a business publication would provide a competitive analysis of one segment's satisfaction with frame relay services. Under this example, a market research firm could call 100 telecommunications managers at universities and ask whether their networks use frame relay. If so, they can then ask whom the supplier is, approximately what they pay, what they like about the service, and what does not meet their needs. The probability is that some of the customers surveyed will be your own, and they will provide direct feedback to you about your own service, and indirectly about your market position. Respondents are most comfortable responding anonymously to a survey, so this method has some limitations as the best way to get feedback about specific customers. Still, a survey of this kind can provide information about your products and how they compare to other products in the marketplace.

Many types of analysis can emerge from a competitive intelligence program. Companies can derive useful information regarding their strengths and weaknesses relative to their competitors. They can conduct due diligence about acquisition targets, merger partners, or potential suppliers. Competitive intelligence programs can support reengineering efforts because they will furnish information about competitors' business processes or business metrics that can be compared to one's own performance.

Analysis of other industries can provide a preview into planned developments in one's own industry. Local service providers have learned some of what to expect from the deregulation of long-distance markets, and many have already reacted in preparation for the eventual battles.

According to The Futures Group, about 17% of respondents to their survey did not believe that competitive intelligence techniques had been used against them. Some companies have made the effort to detect competitors' intelligence efforts. When they think to ask about it, these companies often learn about strangers calling for information but not identifying themselves, people counting cars in the employee lot, and other seemingly innocuous behavior that might gain competitive intelligence. Receptionists and guards can alert management to competitors' efforts as they detect them.

Employees in general should be informed that the information they use in their everyday work is of interest to competitors, at the risk of their own company's success. Employee orientation should include a short briefing on recognizing counterintelligence efforts, and a clear understanding that company information is private and confidential.

Competitive intelligence programs do not sprout by themselves. They require planning, commitment from management, and the tools to be successful. The first step is to assign a budget to the function of competitive intelligence and assign at least one person to be responsible for the program. Competitive intelligence, like many planning functions, will be accomplished only if the individual responsible does not have operational responsibilities and urgent timelines in unrelated matters. The tools the competitive intelligence professional needs include the following:

- Access to online data (including for-fee databases);

- Training in competitive intelligence techniques;

- Compliance training in legal risks and requirements and the company's code of conduct;

- Information systems support (such as network-based knowledge management software, or internal e-mail templates and surveys);

- Routine consultations with senior management.

The competitive analysis professional needs to have capabilities beyond a general understanding of the industry and market. Among other skills, this individual must be familiar with database research techniques and have communications and interviewing competency. A skilled analyst knows which data sources are reliable and has the analytical expertise to draw sound conclusions from incomplete data. Last, because the process is by nature an investigative one, the individual should be willing to work in a uniquely unstructured environment.

Management's role does not end with budgeting for the competitive intelligence function. Without the resolve to develop action plans and ensure that responsive actions are completed, management will not realize the value of the program.

SELF-ASSESSMENT—COMPETITIVE INTELLIGENCE

The following questions will assist telecommunications marketers in determining whether their present competitive intelligence processes are adequate.

- Does your company maintain a formal competitive intelligence program? If not, would you benefit from starting one?

- Do your employees know what to do if they obtain valuable competitive information? Do your employees know to conduct themselves ethically?

- Do you purchase and use competitors' products, including the customer service and other support?

- What do you know about the plans of suppliers or major customers?

- Are your products very similar or somewhat different from those of your competitors? If they have favorable differences, how close is your competitor to overcoming those differences? Is your product (or theirs) protected by patent or other barriers to entry?

- Does your company know whether others are conducting competitive intelligence against you? What have you done to protect your company's private information? Have employees been trained in counterintelligence techniques?

References

[1]   Piirto Heath, Rebecca, "The Competitive Edge," *American Demographics*, Marketing Tools Supplement, pp. 66–73.

[2]   Greenberg, Ilan, "Spy vs. Spy," *The Red Herring*, Issue 44.

5

Market Segmentation

The rise of segmentation

The concept of market segmentation came about when companies realized that markets are heterogeneous. Customers are not alike, so products should not be alike either. Companies then recognized that they could increase their overall level of sales by formulating different marketing strategies for subsets of the market.

A market segment is a group of customers and potential customers with similar needs and similar buying patterns. A market segment is also characterized by the fact that its needs and buying patterns differ from those of the general universe of buyers.

Customers do not have to know whether they fall under a particular segment. In fact, customers are not limited to a single segment. At work, the telecommunications manager at a Fortune 500 company falls within a segment of high-volume customers. At home, the same person may have one or two telephone lines, with needs that match those of a particular consumer segment. As a large enterprise buyer, the customer might demand a high level of customer service and be willing to pay for the most reliable service possible. As a consumer, the customer may want only a basic package for each of the lines and would not be willing to pay premium fees for enhanced services or to ensure reliable service.

Even customers in the same role can fall under different segments. A single consumer could have the characteristics of the early adopter (those that are quick to embrace new technologies) and have the characteristics of the high-usage long-distance user. Segments can get more complicated. One program could target high-usage early adopters, and another could target low-usage early adopters.

Segmentation has become more evident in virtually every industry due to several trends. First, *economies of scale* are less critical than they were at the dawn of the industrial revolution. Henry Ford's "any color you want, as long as it's black" reflected early industrial constraints on automobile production. This statement also demonstrates clearly the homogeneity of products in the early part of the twentieth century. Economies of scale drove many management decisions. Monopoly telecommunications service providers offered few choices to keep overall costs low. Customers simply could not have the customized products and services they desired. In more recent decades, computer technology and other advances have reduced the impact of large scale on unit costs. Technology has also facilitated the science of logistics, which reduces the overall costs of production and distribution and provides a cushion for new costs introduced by segmentation.

Globalization has introduced new cultures, new products, and new needs to the marketplace. The worldwide availability of products has increased competition, which leads producers toward segmenting markets in search of sales. Accessibility to foreign products and cultures also increases the information available to customers and heightens their taste for variety in their purchases.

Changing *demographics* foster segmentation. The average customer is more educated and wealthier and has access to more product information than in the past. If a customized product is available, the customer has an excellent chance of learning about it.

Perhaps most important, advanced *information technology* has enabled companies to conduct the research necessary to derive meaningful market segments and make accurate measurements of their buying behavior. The amount of data required in this effort goes well beyond the capability of manual computations. Only the most advanced computers and sophisticated mathematical analysis can make this level of segmentation both possible and cost-effective.

The marketing mix

Market segmentation involves redesigning the *marketing mix* to meet the specific needs of a subset of the customer base. The marketing mix consists of four variables: product, pricing, promotion, and place (distribution).

The *product* is more than the physical product or service being sold. Attributes of the product include its quality, the customer care, and the reputation of the provider. Many attributes of the product can be tailored to the needs of the customer segment without altering production capabilities.

The *pricing* can be varied to meet the needs of each segment of the market served. Prices can be usage-based or fixed, monthly or annual. Pricing to meet the needs of market segments is worthwhile because different segments will have different price tolerances. If price is very important to a segment, customers will leave if they become aware of competitive products at a lower price. On the other hand, if buyers in the segment would pay a higher price than the average customer would, it would be foolish to lower the price and more prudent to add features to the product to retain the customer.

Distribution, or place, refers to the methods by which customers can buy the product. These decisions involve whether retail is an appropriate channel to reach the segment, or if direct marketing techniques such as telemarketing or mail order are more effective. Some market segments, such as large business, require direct sales. The costs and the effectiveness of channels vary. Segmentation enables the seller to distribute products through those channels that provide the most sales at the least cost.

Promotion refers to the communications between the provider and customer. Television advertising is quite effective to reach a very large segment but expensive. Moreover, it is extremely difficult to convey the marketing message in the limited time provided by an advertisement. To reach a more concentrated segment, perhaps a very technical industry group, other venues are more targeted and less costly. Trade publications or trade shows would constitute a better choice for very technical offerings. Promotions also include temporary product offerings to stimulate sales to a segment, such as unique bundles, introductory pricing, or tie-ins to other services valued by the segment.

Segmentation in regulated markets

Market segmentation is not new for the telecommunications market. Under regulation, markets are segmented, but for rate-making rather than marketing purposes. A primary form of segmentation occurs between monopoly telecommunications services sold to businesses versus those sold to consumers. Identical services (such as a single unlimited local line) offered to businesses or consumers are priced using different rate schedules. This is true although the cost of providing those services is equal—and despite the fact that it is often less expensive to provide to the business user, who most often pays more. The price of access in one state can vary considerably from the rate in an adjacent county in another state. Regulated monopoly pricing is grounded in social policy, not cost or market demand.

Regulated telecommunications providers also segment by product, often to meet separate subsidiary requirements or other regulatory provisions. Regulated telecommunications providers are frequently required to sell customer premises equipment (telephones) through a separate subsidiary. This is also true in some jurisdictions of competitive services such as mobile telecommunications. These organizational anomalies can result in the creation of multiple sales channels for very similar products and different avenues for customer service. Wireless customers of RBOCs purchase the two types of service from separate channels. They commonly receive separate bills for wireline and wireless service and have two avenues for customer service.

Regulated telecommunications providers do practice segmentation for reasons not involving regulatory mandates. Some segmentation is conducted to increase high-profit sales. Even in a regulated monopoly, some large business customers can purchase services through account managers and other direct and indirect sales channels not available to other customer groups. Low-profit consumers visit company offices or call service representatives. Although some segmentation is in evidence, monopoly providers generally lag behind competitive markets in their use of segmentation techniques. Deregulation is likely to spark more sophisticated and more practical marketing strategies.

How deregulation gives rise to segmentation

Under a monopoly environment, products and services need to meet customer needs, but there is little incentive for the provider to individualize services beyond these basic requirements. In the aggregate, market segments and the highly individualized services they bring about will add to overall cost, which adversely affects prices. After all, the customized services are generally desired by high-volume or business customers that are already paying subsidies to sustain universal service. Beneath a monopoly umbrella, it does not make economic sense to differentiate service offerings excessively. Without competitors that offer more customized products, there is little pressure to take the risks of a segmentation strategy.

Deregulation promotes the segmentation of markets for several reasons. First, deregulation raises the intensity of competition, which pressures all companies to offer products that customers view as superior. Providers need to find ways to differentiate products by appealing to individual needs.

Second, deregulation encourages smaller providers that cannot afford to serve everyone to target the needs of a market niche. Companies that do not have the resources or the regulatory mandate to serve all segments will select the segments that will provide them with the most profit. Their targeted customers will appreciate the custom services and will choose them above the more generalized offerings of an all-purpose provider. As a competitive response, all providers will eventually have no choice but to offer more targeted offerings.

Third, competition places enormous pressure on the prices of equivalent products. When products are homogeneous, the seller with the lowest price will always gain market share. Price-based competition is not the most desirable market strategy. The only way sellers can compete on a basis other than price is to create products that more closely meet customer desires. This can most effectively be accomplished through segmentation.

Why segment markets?

Notwithstanding the potential need for segmentation as a survival strategy, there are several reasons that a segmentation strategy will lead to increased profitability. Segmentation can improve profitability by increasing revenues from the most valuable customers. The closer the segment description comes to matching the customer's needs, the more likely it is that telecommunications providers can offer products that cannot be paralleled by competitors. Thus, segmentation increases market share. Moreover, customers demonstrate less price sensitivity and more loyalty when a product is a close fit to their needs. Consequently, segmentation increases profits and reduces customer loss.

Since its deregulation, long-distance service appears to have high price elasticity. With all other things equal, the company that offers the lowest price gains market share. Without any segmentation strategy, a long-distance provider may conclude that revenues would increase the most if it lowered the price of long-distance service to all customers. With a proper segmentation strategy, the provider could choose the most price-sensitive segments and offer to lower only the rates in those segments. Share in those segments would rise, without unnecessarily sacrificing profits in the segments that are not as price-sensitive.

This analysis assumes that there is some way to reduce one segment's price without a commensurate drop in all segments. Offering low prices for calls made after 10 P.M. is an example of how to accomplish this. This strategy succeeds because the price-sensitive callers can choose to wait until calls are inexpensive, and the business caller is unable to benefit from the new pricing.

Segmentation can enhance providers' profitability by eliminating unnecessary costs in designated segments. One cost that deserves scrutiny is distribution. For example, a market segmentation analysis discovers that a certain distribution channel or advertising venue is drawing most customers from a segment, and another is drawing none. Alternatively, customers in a geographical area might be responding to telemarketing calls but not responding to a full-page newspaper ad. If the less productive venue is not essential to another segment's strategy, its costs can be reduced or eliminated. Without segmentation analysis, each distribution channel would appear to be equally effective.

Segmentation often improves the company's revenue forecasting capabilities. Besides helping financial departments produce better reports, better forecasting will result in having the capacity to meet market demand, without creating overcapacity, thus reducing operations costs. Several U.S. local wireline companies were caught in this situation when they did not anticipate the rapid popularity of the Internet. Consumers made two rapid market shifts: They purchased additional lines, and they used their access lines for long Internet sessions. This caused two serious problems. First, operations divisions were unprepared for the workload to install the new lines. Second, the long online sessions were straining central switching office resources, which had been configured in anticipation of normal voice traffic.

For the largest providers, segmenting markets does not necessarily require selecting a subset of markets to serve and specifically avoiding others, nor does segmentation necessarily demand a reduction in the scope of services that a company offers. The company can still allocate each customer in the entire customer universe to one or several of its designated segments. On the other hand, segmentation and its subsequent analysis can provide data supporting a decision to reduce scope. The provider could find that a specific segment is given to unprofitable churn, is too costly to acquire, or creates a disproportionate amount of uncollected bills. If there are no excessive barriers to exit, the telecommunications provider could decide based on this analysis to discontinue serving the segment.

Types of segmentation

Segments have to meet several criteria to be useful in developing services for specific markets.

- The group needs to be large enough to justify the effort expended in targeting them.

- Members of the segment must be able to be differentiated from other customers.

- The segment must be accessible in a cost-effective manner.

- Sales channels must be adaptable to meet the needs of segments targeted.

- Customers must be able to perceive differences between products.

Companies moving from a regulated to a deregulated marketplace need to revisit their segmentation strategies to ensure that they are optimized. Business customers and consumers represent market segments that have been treated differently in the past. In the future, business customers will probably continue to consent to pay a premium for service, but competition in this market will put pressure on providers to offer more value in exchange for the higher prices. Subsets of these distinctions will emerge. Business customers in offices have different needs from those in home-based businesses. Families' telecommunications requirements differ from those of the elderly. Families' telecommunications requirements change over time, first when they have infants or small children, then when teenagers live at home, and again when they have children at college.

Other segments, those resulting from monopoly or interim segmentation strategies created by the migration to deregulation, are less likely to survive. Product segmentation, such as wireline and wireless service, has already shown signs that it could disappear in the next decade. The local and long-distance split is nearly certain to change significantly. Other product segmentation strategies that are presently visible in the market, such as voice/data/video or local service versus local Internet access, are likely to change dramatically. Many of these product segmentation strategies could have less meaning if networks move from their current architecture to IP architecture, as is widely anticipated.

Segments vary by industry, especially telecommunications-intensive industries such as airlines, financial services, and teleservices. Segments vary between consumer and business markets and between wholesale and retail markets. The following descriptions of potential segmentation strategies can be a starting point in a segmentation analysis.

In *demographic segmentation*, buyers are grouped according to personal characteristics they share. In consumer markets, these characteristics could include age, gender, nationality, household attributes, income level, and other traits that affect the type or amount of telecommunications services they use. Demographic segmentation is probably most effective in consumer markets. BellSouth has targeted service bundles toward teenagers and college students. Internet service providers have made some effort to segment demographic groups.

One significant demographic variable could be highly meaningful with a targeting strategy—the urban versus the rural information services user. Urban users can be served at a lower cost, and rural telecommunications is high-cost. In a highly competitive market, urban users will have many choices. Rural customers will likely have fewer choices than their urban counterparts.

Under regulation, for the most part, both segments receive similar services at similar prices. Rural customers, who have less access to urban distribution facilities and other resources, will be most dependent on online resources. On the other hand, they will benefit more from telecommunications services than their urban counterparts with greater physical access to stores and information. Rural customers' willingness to pay is unknown, and their pricing legacy and expectations are most divergent from actual costs. Still, based on the benefits available to rural customers, this segment might represent a significant market opportunity to the provider that makes the right offer.

Geographic segmentation groups customers by their physical location. In a regulated market, providers segment geographically because of the limitations of their franchised territories and the specific desires of regulators. Competitive providers will need to examine the legacies of regulation carefully. The line drawn to ensure that the firehouse across town is a local call could still be in place 10 years after the firehouse is moved. In present deregulated markets, facilities-based wireless carriers or Internet access providers could segment geographically in areas where they operate facilities. Smaller providers have used geographical segmentation to focus their facilities and their other resources on a service area. Internet providers that serve consumers have found that they can differentiate their product by providing local news, shopping, and other information of local interest. Their product becomes more valuable to the local customer because the national providers have a limited ability to provide local content in every area they serve. As long as the customer does not value the differentiating attributes of the national providers, such as worldwide local access points, the small local ISP will win the customer's business.

Behavioral segmentation defines buyers through the ways they use the service. A segment might group all of the customers who use the product in the same way, such as all international callers to China. In consumer markets, behavioral segmentation can target sports fans or hobbyists. Behavioral segmentation identifies the segment's common incentives and develops loyalty programs and service bundles to meet the group's specific desires. The company that targets callers to China could offer discounted rates to China or monthly promotions that raffle a vacation in China.

In business markets, *organizational segmentation* seeks vertical markets, such as an industry group. As the telecommunications industry matures, providers will develop solutions that meet the complex needs of industries such as financial services, education, or health care. Some local telecommunications providers are investing in telemedicine to attract the health care industry. As markets become more competitive and more specialized, companies in an industry tend to gravitate toward telecommunications service providers that offer solutions to their specific industry's concerns.

Direct marketers sell products through catalogs, direct mail advertising, or television offers with a toll-free number. It would be feasible, for example, for a telecommunications provider to offer a package of services and products to a direct marketer. The package might include inbound toll-free services, computer/telephony integration services (that enable sales representatives to view customer account histories from caller ID information), and Internet video services. Video services would allow customers to view and interact with service representatives from their computers. Perhaps the telecommunications provider could add a software package to the service bundle that helps the customer to track sales effectiveness. The direct marketer would choose to purchase this package to simplify the procurement process, to ensure that all the components are integrated, and to work with a single vendor.

Analysis techniques

Companies can either define market segments based upon their customer knowledge and experience or use statistical techniques to uncover the segments for them. Predefined segments offer an intuitive comfort, but they can be based on inaccurate assumptions or create suboptimal groupings. Technology has made it possible to conduct sophisticated analysis to profile customer records and generate candidate market segments. Several PC- and mainframe-based computer programs are available to assist marketers in evaluating the market segments in their customer base.

Cluster analysis groups customers based on their similarities to each other and their differences from those who are not in the segment. Cluster analysis can determine the profile of the typical purchaser of enhanced services or identify the customer who is likely to leave. When companies serve millions of customers, and maintain large databases of information about each one, it is virtually impossible to identify these groups without statistical support such as cluster analysis. One largely unexploited benefit already held by telecommunications providers is the enormous amount of historical usage information they collect about their customers' interaction with the service. Cluster analysis sorts this data into meaningful categories.

The *chi-squared automated interaction detector* (CHAID) technique is useful to create market segments with statistically significant differences. The resulting segments can have several variables, but all variables do not necessarily appear in all segments. The results of a CHAID analysis could be, "We have identified the segments that are 80% or more likely to spend more than $100 per month. They include couples with children, singles with a college degree, or childless couples who take more than five vacations a year." From this analysis, promotional efforts could target prospects with the defined attributes who are not presently customers, or target the remaining 20% of existing customers in the segment and work to increase their usage.

Correspondence analysis can take two variables and graph their relationship to each other. Suppose that one variable consists of the various services offered by the telecommunications provider. The second variable then lists several attributes that the customer associates with the service. Perhaps the attributes are value, reliability, customer service, and usability. The resulting analysis would chart the characteristics most associated with each of the services. This analysis can serve as a foundation for a segmentation strategy that targets certain services to customers who value the attributes of the service. If an analysis of this form demonstrated that customers associate a certain service with a high level of reliability, this could inspire an advertising campaign to attract customers who are seeking high reliability.

The costs of segmentation

Segmentation is not free. In addition to the management planning and analysis costs of market segmentation, any new revenues resulting from segmentation will need to offset some other costs that the process incurs.

A segmentation strategy can result in the development and production of new services. This decision will probably reduce the economies of scale characteristic of a single-product operation, although it would not eliminate them. Product design and production operations will each add to the cost of creating the targeted service.

Targeting markets often implies new advertising campaigns, distribution tactics, or promotions. Such efforts require development, administration, and measurement of their effectiveness. Fortunately, promotional campaigns targeting certain users are less expensive than campaigns targeting all users. Still, the overall promotional budget is likely to increase.

For services that involve physical components, inventories will rise. Even when the physical service does not involve additional inventories, costs can rise. When a telecommunications service that includes complex usage instructions is developed for a specific segment, the manuals will add to inventories. If different segments require alternate hardware configurations, inventories must carry an adequate amount of materials.

Management attention will also represent an added cost of segmentation. Often, companies engage a market manager who monitors the segment, evaluates current programs and market share, and ensures that the service line continues to meet the needs of the segment.

Segmentation strategies in telecommunications

Although the competitive telecommunications market is still in its infancy, there are indications that telecommunications providers view segmentation as important. Several of the RBOCs and all of the major long-distance carriers have created divisions directed specifically at the needs of small business. Some providers support divisions or at least account management covering vertical industry groups such as government. Most of the LECs offer targeted marketing to resellers of their services. Some have embraced resale as a potentially gigantic and enduring market segment; others have been criticized on the basis that their commitment to resellers is grounded mostly in regulatory mandates. Whatever their motivation, the organizations serving resellers often include separate customer care services and business support.

Airline deregulation is sometimes viewed as a model for telecommunications deregulation. This industry offers some lessons in market segmentation. Airlines routinely segment between business travelers and consumers. The most apparent differences in the two air travel segments are price sensitivity and volume, and they occur between business travelers and vacationers. Business travelers have a stronger tolerance for higher prices. This is partly because of the value of the trip to the sponsoring company and partly because the flyer often does not personally pay the ticket price. The volume traveler is desired by airlines and courted with frequent flyer miles, which increase the flyer's personal benefits; class upgrades; and airline clubs, which increase the comfort of the passenger while traveling. All of these techniques are intended to increase the loyalty of the passenger, and they work well. Many of these characteristics are also true of the business telecommunications user. Users who make the purchase decisions do not necessarily pay the bill. They have immediate communications needs that cannot be met by making the call during off-peak hours. They are willing to pay a premium price for higher quality services. Telecommunications providers are responding to the needs of high-volume users by modeling the airlines' marketing techniques, including awards programs for airline frequent flyer miles.

Segment tactics

Once the provider has identified the differences in market segments, how does it meet the segment's very specific needs? The marketing mix provides some starting points. The provider can address the buyer's decision criteria by altering the product, pricing, promotion, or distribution of the service. Furthermore, when the buyer's needs are discovered to be homogeneous, the provider can choose not to change the elements of the marketing mix. Therefore, segmentation can result in differentiated, undifferentiated, and niche marketing.

The most valuable outcome of market segmentation is the ability to develop and package differentiated services that meet the specific needs of an identified market segment. In the telecommunications market, *differentiated marketing* can build on any element of the marketing mix.

The basis for differentiation can be the nature of the product itself. Wireless providers offer thousands of bundled minutes with a large minimum and a low per-minute charge for high-usage customers and flat rate service with a high-usage charge to consumers seeking mobile communications only for emergencies.

Pricing has traditionally served as a common means of differentiation and probably will continue to do so. Business and residential pricing for essentially the same product was historically disproportionate to the cost of providing service. While the business market is certain to become highly competitive—including on price—it is possible that the segment will continue to pay a premium for service, perhaps in exchange for some other product feature. Business air travelers routinely pay fares well in excess of consumers, in exchange for travel flexibility. Business telecommunications users have already demonstrated a willingness to pay and less price sensitivity than consumers have. The successful telecommunications provider of the future will find a way to hold both the customer and the premium price. Market researcher Technology Resources observed that about two-thirds of survey participants would pay an additional $5 per month to add one telephone number anywhere to the flat rate service presently offered [1]. Competitive providers will need to be creative in designing services that entice customers with varied needs.

Services are differentiated through promotion. A prepaid calling card with a certain value and a certain per-minute rate can be marketed to several groups by emphasizing some benefits and ignoring others. The merchandise display in a distressed urban area could accentuate the low rates and the ability to make calls at public telephones without holding a large amount of change. To a business manager, the promotion would include the ability to limit the amount of an employee's out-of-town calling. The exact same card can be differentiated in a multitude of customer segments, if the provider knows the segment's buying needs.

The distribution of the product can also be varied to meet the needs of different customer segments. Sales and distribution resources can be concentrated at the places where the target customers are expected to be.

It seems contradictory that a company that bothers to segment its markets would then decide to market identical products across all segments. Nonetheless, *undifferentiated marketing* is the proper choice when segmentation research concludes that certain product attributes do not need to be differentiated.

Therefore, undifferentiated marketing is appropriate in two situations. When the entire market's needs are homogeneous in a certain area, segmentation is unnecessary. In the telecommunications market, this can occur when the entire industry voluntarily conforms to standards. Dialing 911 for emergency assistance is a standard. Few marketers would entertain the notion of changing the dialing sequence as a targeting strategy, even if they were permitted to do so.

The second occasion in which undifferentiated marketing is appropriate is when the telecommunications provider chooses to compete on price and sells the service as a commodity. For those customers whose primary buying criterion is price, the commodity nature of the product will appeal to the customer's perception of value. This presumes, of course, that the provider can create genuine cost savings in producing the commodity service, and pass along these savings to customers.

Niche marketing resembles differentiated marketing in that the marketing mix is directed at a subset of customers. While differentiated marketing is most often practiced by the largest competitors, serving several or many segments, niche marketing is available to very small competitors. Niche marketing (or *concentrated marketing*) targets a single market segment. Examples of niche providers include telecommunications service providers offering satellite telecommunications to one particular trade only, or those providing prepaid phone cards only in Spanish with service only to Cuba. These providers have the opportunity to excel in the eyes of their target market and can serve these markets without making investments on the same scale as those of their more generalized competitors.

These niche providers are of critical strategic interest to the largest telecommunications companies. Separately, their presence forces the giants to develop a strategy for the markets served by the most successful niche providers. In the aggregate, niche providers can constitute a considerable portion of the market as a whole.

A nondifferentiated strategy may not draw customers. A differentiated strategy might be cost-prohibitive for a market of niche size. Exiting all niche markets will impair scale economies and reduce overall market share. Reselling to the niche providers could be a worthwhile approach. In any case, all telecommunications providers will be very interested in segmenting their markets and understanding the dynamics of each segment.

SELF-ASSESSMENT—MARKET SEGMENTATION

Responses to the following questions will help telecommunications marketers determine the success of their present segmentation strategies.

- Does your company segment its customers? Do the present segmentation strategies represent customer needs or legacy groupings?

- Are the products differentiated by segment? What about the other elements of the marketing mix?

- What analysis by segment do you conduct regularly? Is each segment required to be profitable? Is market research conducted by segment?

- Which segments are most valuable? What investments have you made to protect these important customer groups?

- Are niche providers a threat to your segmentation strategy? What actions are required to manage these threats?

- Is your infrastructure capable of further market segmentation?

Reference

[1] Page, Rodney, and Jim Spann, "Think Different", *Telephony*, Vol. 235, No. 16, pp. 68–74.

Part III

Distribution Strategies

6

Channel Strategies

Channel alternatives

A *channel* (or *distribution channel*) is the organized support structure that enables a provider to reach customers and enables the customer to find the desired provider. A virtually unlimited variety of distribution channels can be created using combinations of distributors and intermediaries such as wholesalers, agents, resellers, and retailers.

In most industries, companies select the channels they want to serve and then limit their business to that choice. In a monopoly telecommunications industry, one franchised company generally dominates all of the distribution channels in each assigned territory. Historically, it was neither necessary nor desirable to limit the channels or the business as a whole.

Telecommunications providers forced to limit the scope of their businesses are very likely to focus their strategy on a single distribution channel. Ideally, as the industry becomes more competitive, telecommunications providers are most likely to gravitate to the distribution channel or channels in which they will enjoy the most success.

As local service is on the verge of deregulation in the United States, companies are now facing the requirement to design effective channels and, in many cases, to commit to fewer channels than in the past. The process will be difficult and momentous for the largest carriers. In the past, covering every channel gave telecommunications providers the opportunity to set price structures that met universal service objectives. As products will need to be self-supporting, so will channels.

Selecting a channel for the new telecommunications marketplace will define who the customer is. In a monopoly, the customer base includes virtually everyone. In a competitive market, targeting each customer is simply not feasible.

Developing channel strategies requires decisions about whether to build facilities or resell telecommunications services provided by a different facilities owner. Building facilities requires a large initial investment, but it enables a company to control most aspects of the service it provides and often provides high profits when these owned facilities are efficiently utilized. Reselling the facilities of another requires less investment but introduces intercompany relationships, less control over one's product, and a lower profit margin in an efficient market.

One of the complicating features of the telecommunications industry is that the costs of the network are disproportionately resident in the "last mile," the cabling between the local switching office and the user's premises. The potential entrants in the local service market are not eager to absorb the expenditures required to build duplicate facilities to connect every telephone user to the network. Therefore, they lean toward reselling services until the competitive market has been established. On the other hand, the incumbent companies, which currently span multiple channels, are reluctant to divide their businesses in the turmoil of deregulation. Other factors, not related to marketing strategy, also affect channel strategy decisions.

Vertical integration describes a business that spans several channels, such as wholesale and retail. Although monopoly providers are vertically integrated and successful, they enjoy advantages that will be unavailable to competitive market participants. The following list demonstrates the protections offered by monopoly to the vertically integrated company. In a deregulated marketplace, the absence of these advantages will be weaknesses for those that choose total vertical integration.

- *Monopoly providers have access to inexpensive, low-risk capital.* They can estimate accurately demands on infrastructure. Investors apply low risk (and low interest rates) because revenues are designed to recover investment. Competitive companies need to demonstrate to investors and lenders that the borrowed capital will be recovered. Thus, interest rates will be higher, and capital will be less available in a competitive market.

- *Monopolists have captive customers.* A monopoly skilled at only wholesale or only retail, or neither, will not lose customers to competitors. A competitive provider, similarly unskilled, will fail.

- *Customers seek out the supplier in a regulated market.* In a competitive marketplace, the provider that waits for a customer to approach will lose the customer to competitors that are more active.

- *Regulated monopolies have limited requirements to expand their businesses.* Competitive companies need to expand into any areas that will add to profitability. Concentration in a single channel increases the probability of success.

- *Monopolists can control their investment in infrastructure.* If a network node is usable and not fully depreciated, a monopolist can avoid replacing it. In a competitive market, competitors will seize the opportunity to capture customers with improved technology.

- *In the vertically integrated monopoly, profitable channels can compensate for unprofitable channels.* A competitive telecommunications provider needs to be world-class in every aspect of the business, or exit.

The transition to deregulation poses challenges to incumbents and new entrants because of the new channel structures. In markets undergoing deregulation, new entrants need to buy services from the incumbent provider. Incumbents resent the need to sell services, often at a discounted price, which simply results in a reduction of their own market share. The new competitors often suspect the incumbent's motivations and operational support. Information systems and business processes are new, untested, and conducted by unskilled personnel from all providers.

Every product category has undergone its own pain caused by channel clashes and competition. Pay phone providers had no alternative other than to purchase special access lines with complex measurement and pricing requirements from the incumbent LEC. The local carrier was required to provide the line and compete against this wholesale customer for the location-owner retail customer. In some jurisdictions, the competitor could charge more for a call at a public telephone than could the incumbent provider.

When long-distance was deregulated, IXCs required connections from local service providers to fulfill their customer orders. This is again a problematic issue as local service undergoes deregulation. Local service competitors complain about the priorities assigned to their installations, and the incumbent carriers respond that their systems meet the requirements mandated by regulators. In general, deregulation is an unpleasant but necessary period for providers and customers alike.

After deregulation, the companies that commit to a single channel are more likely to invest in a support infrastructure for their own channel, including sophisticated information systems and employee development. Marketing divisions will need to work closely with other corporate functions to ensure that the company's infrastructure can support a concentrated focus.

In the wholesale market, this infrastructure includes sophisticated operations support systems for processing service requests. Wholesalers' customers are providers, too, and their information needs are very different from the needs of end users. Wholesalers will need to develop sales organizations that match the knowledge of their customers and that are skilled at large account management. Serving multiple providers, they will require information systems and internal processes that protect their customers' data and trade secrets.

Wholesalers will undoubtedly seek to compete through their operations support systems and information management capabilities, especially if the network services alone resemble a commodity business. They can offer their customers management reporting or decision support data, capacity planning services, or other services that leverage the network management knowledge of the wholesaler to the benefit of the customer. Wholesalers could also try to enhance the actual telecommunications services offered with software-based features or innovative packaging, or offer engineering and technical support concerning the switches or other network components held by the retailer.

In retail markets, companies will develop stronger capabilities in targeted marketing, perhaps including industry expertise in niche markets. Information systems will need to accommodate customer care and telemarketing. Retailers will make large investments in advertising and learn from successful and unsuccessful campaigns. Retailers without facilities investments will be able to enter and exit markets or switch wholesale providers to meet market requirements.

Both wholesalers and retailers will need to understand their own cost structures and learn about the very specific needs of their customers. Some companies will continue to serve all channels and will make lesser investments in their supporting business processes. All other things being equal, the company that is focused on a single channel will defeat the company that is scattered among many markets.

The *vertical dimension* of the channel describes how many intermediaries are involved between the manufacturer and the customers. For a roadside fruit stand, the producer of the merchandise has access to the customer, and intermediary channels are unnecessary. Most consumers do not buy fruit from the roadside stand, though. The simple purchase of an apple can involve a wholesaler, a distributor, and a retailer.

Certain areas of the telecommunications service business are more direct and require fewer levels of distribution. Some of the largest corporations will buy their huge switching systems directly from the manufacturer and use them to provide telecommunications services to employees. Generally, products that are high-value and complex require the shortest vertical channels.

Most areas, though, require many participants in the process before a customer can buy and use telecommunications service. For telecommunications service, the individual customer needs, at a minimum, a service provider and a telephone. Each of these requirements is supported by wholesale and retail channels. There could be other participants in the distribution process as well.

The wholesale channel

Telecommunications wholesalers differ from the traditional characterization of wholesalers. Wholesalers of most products are not manufacturers; they purchase manufactured products in bulk, break the bulk into smaller units, and distribute to retailers. Instead, wholesalers of telecommunications services take manufactured products (switches, cabling, and other facilities) and build the infrastructure to create and support a service to sell to customers. In a sense, the telecommunications wholesaler is indeed a manufacturer of telecommunications services.

The role of any wholesaler is not simply to buy in quantity and sell in smaller quantities. In the more mature nontelecommunications industries, companies serving the wholesale channel have no direct contact with individual customers, but they do whatever they can to create sales, without actually performing the sale. Wholesalers advertise lavishly or provide incentives to customers or retailers. They can offer training in sales or use of the product to the retail channel. Successful wholesalers are not passive.

Telecommunications wholesalers generally own enormous interexchange network facilities and charge their customers for the use of these facilities on a measured or a leased basis. Today, the primary measurement for network usage is time, but wholesalers are experimenting with other units of measure related to the quantity of bandwidth consumed. The telecommunications providers that are already committed to serving wholesale markets are companies such as Williams Communications, often labeled a "carrier's carrier."

Nontelecommunications companies whose infrastructure can be leveraged are interested in entering the industry through the wholesale channel. Energy utilities are expected to provide interexchange facilities on a wholesale basis. Their existing plant includes valuable rights-of-way, and sometimes utilities install and operate fiber networks for their own internal use. Wholesaling is a valid entry strategy for a company that does not need the distractions of dealing with end users. Sprint began as a facilities wholesaler that leveraged rights-of-way; in fact, the name Sprint was based on the acronym of its original parent company, Southern Pacific Railroad. Defense companies, historically a research and development source for the telecommunications industry, are developing commercial opportunities in telecommunications markets [1].

Wholesale companies primarily provide services for others to sell to customers. Interexchange carriers such as MCI and AT&T occasionally resell each other's facilities when their own facilities do not meet either geographical requirements or traffic levels, but they are generally committed to the retail channel.

The wholesaling of local services is anticipated to be an important channel, perhaps more significant than interexchange wholesale is today. Because the local portion of telecommunications service is such a large cost component, competitors are expected to share and resell facilities for the long term more than they would in markets with low fixed costs.

Buyers of wholesaler-supplied services will not be limited to resellers. Wholesalers will be able to serve any high-volume customer with the expertise to act as a retailer and manage the network it buys for its users. A wholesaler's customer base can include multinational corporations and Fortune 500 companies. Technology-dependent vertical markets represent another potential wholesale segment. Industries in this segment include financial services, airlines, aerospace, and the top call centers, all of which depend on telecommunications to run their operations. These companies all maintain internal expertise to support their complex internal networks.

Retail channel overview

Retail describes the direct relationship between a seller and an end user. In most industries, the concept of retail is a positive one, evoking the image of a shopkeeper entrusted to create sales and profits. Unfortunately, in the telecommunications industry, the reseller/retailer was introduced as the new market participant whose intention it was to take customers and profits from the franchised monopolist. Whether justified or not, the incumbent provider was viewed with suspicion by the retailer and the reseller viewed with disdain by the wholesaler. These lingering sentiments form the foundation of the wholesale/retail relationship in telecommunications, unlike most other industries. Eventually the exigencies of a competitive market will create a more realistic relationship between network providers and their retail representatives.

While the retail channel represents the last stop on the vertical dimension of distribution, retail can demonstrate the *horizontal dimension* of the channel. A variety of retailers resides on this level of distribution, and a manufacturer or wholesaler is not limited to one representative in any given category.

Many retail venues are emerging, and telecommunications providers have made overtures toward most of them. Retail venues include stores, superstores, direct marketing, kiosks, and the Internet. Some of the vertically integrated providers are augmenting their own efforts with alternative retailers. Sprint accompanies sales of its PCS service at Radio Shack with a substantial advertising and comarketing campaign [2]. Vertically integrated wireless providers support company-owned stores with agents in their own stores and kiosks in superstores. Arranging for additional retail venues enables telecommunications providers to bundle their offerings with other telecommunications services that are beyond their own product lines, such as prepaid telecommunications, wireline hardware or services, or Internet access.

BellSouth created a retail product that it calls "lifestyle packages," directed at demographic segments and sold at various retail outlets [3]. While these packages contain only instructions for getting a telephone line and several bonus products, their innovative packaging represents the recognition by BellSouth that telecommunications providers will need to seek out the customer in the deregulated marketplace.

Migrating services to the retail channel will probably require telecommunications providers to tailor the product to the channel and establish sales procedures. Because the retail consumer often evaluates the product with little or no on-site assistance at all from the service provider, both the packaging and the product itself need to be very simple to understand. In retail, sales personnel need to know enough about the product to provide more insight than the packaging, so retailers need training. The behavior of the sales personnel will directly affect the customer's perception of the product and the brand.

Direct marketing can be carried out through catalogs, through outbound or inbound telemarketing, and alternatively through the Internet. Sales representatives who take calls need to be knowledgeable and accurate, and the printed material and Web site need to anticipate customer questions and create interest and excitement for the product. In the future, the Internet buyer should be able to order the service and expect its provisioning while on the Web. (The potential of the Internet as an alternate channel is discussed in Chapter 18.)

Each venue has advantages and associated costs. Telemarketing is an inexpensive channel that has apparently been successful in the long-distance market, but some retail customers will prefer face-to-face service. Other users prefer to learn about their alternatives with the new research capabilities available to them through the Internet. Retailers will need to test the alternative venues to arrive at the optimal mix.

Facilities-based retail

At present, the vast majority of local telecommunications services are offered by facilities-based retailers, the thousands of private and government-owned telephone companies in the world.

Today's nonincumbent telecommunications providers generally aspire to become facilities-based providers in new markets. IXCs want to own facilities and sell a full line of services to customers. Incumbent local exchange providers have expressed a desire to build interexchange networks and local facilities outside of their existing territories. Facilities-based retail is the stated goal of most industry participants. In the aggregate, these stated goals portend an enormous increase in local and interexchange capacity.

Some locally based energy utilities will choose to leverage both their rights-of-way and their customer base and enter the retail channel primarily to serve consumers. A few municipalities have also stated their intention to offer local services to consumers.

If long-distance and nontelecommunications businesses are any indication, competitors will build their first facilities in areas that can show immediate returns. While the preponderance of local service is priced under its cost, there are pockets of profitability. Large businesses and office campuses can connect via high-speed facilities directly to an IXC or any provider that connects to the public network. Apartment complexes and high-rises offer dozens or hundreds of customers in a single location.

Not surprisingly, many competitive LECs are starting their market penetration in these areas.

AT&T has augmented its buildout plans with acquisitions of local facilities providers TCG and TCI. In addition, AT&T has priced its wireless services very aggressively to compete with wireline local service. Its stated strategy clearly anticipates that the company will own the entire connection between itself and its customer.

Even within the facilities-based retail category, hybrid solutions will undoubtedly emerge. Besides the companies that are facilities-based in some markets and resellers in others, some companies will choose to resell certain facilities and build others. One example is to build local switching facilities but not access plant to the customer's site. This hybrid enables the provider to lease the commodity elements of the network but add value in the form of software enhancement to the services offered to customers.

Resale/retail

This channel represents the provider that owns very few or no facilities. The provider adds value to the product through packaging, marketing, and customer support.

Most of the major IXCs anticipate entering some markets with facilities-based service and others through resale. Companies such as AT&T, MCI, and LCI (since acquired by Qwest) have made similar statements concerning their market entry intentions. Their apparent strategy is to resell services until enough customers create the required critical mass, then build facilities to control costs and service. The progression that begins with resale moves to purchase of unbundled services (often available at a flatter rate than simple resale), then to facilities-based retail, where higher fixed costs translate to higher profit when volume is sufficient.

It is widely theorized that resale of another's facilities is simply an entry strategy and that companies cannot earn enough to operate this way in the long term. On the other hand, there are several reasons to challenge this assumption.

- *The ultimate mathematics of wholesale and retail are unknown.* Because the present resale costs and customer price structures are very preliminary (and somewhat arbitrary, as they have been developed primarily by regulators), it is premature to decide whether a business can survive on today's margins.

- *The success factors in local service are unknown.* The importance of marketing, bundling, customer care, and other retail services may have been underestimated as a differentiator in markets until now.

- *The structure of a competitive industry is still unknown.* Most large telecommunications providers are extremely vertically integrated; that is, they span several channels. When some of the larger companies divide their channel businesses, either through organizational boundaries or through spin-offs, the resale sector will be very attractive to their wholesale divisions.

- *The local markets that exhibit overcapacity will foster a resale business and competitive pricing.* In long-distance markets, as soon as new companies offered wholesale interexchange facilities, price competition became fierce. While the long-distance and local cost structures are not entirely analogous, in some markets, the second and third facilities-based provider will want to generate usage on its new networks. Resale is an attractive mechanism to utilize overcapacity.

- *Resale is growing in share of the long-distance business.* The thriving resale business in the long-distance market shows no signs of either disappearing or converting to facilities-based retail.

In the long-distance market, resellers are generally smaller and much more numerous than facilities-based carriers. Of all entry strategies, resale is the easiest to launch and the easiest to exit the market as well.

According to the Telecommunications Resellers Association, wireless resale constitutes about 5% of the sales in that market. Market researcher Northern Business Information estimates that resale represents about $6.5 billion of the $95 billion long-distance market [4]. In the United Kingdom, resellers report annual growth rates of 50–100% [5].

There is a more compelling reason to believe that resale, or retail, will be an important channel and perhaps the most visible channel in a mature market. Most other industries that sell to consumers or other individual, low-volume users such as small business have a strong retail channel. Retail is the way most products are sold to the end user. Retailers generally do not create the product they sell, except in regulated industries. Automobiles, consumer products, and travel services are sold by representatives of the manufacturer, not the manufacturer.

For the business buyer, most significant purchases require direct sales. As in consumer businesses, the manufacturer or wholesaler is not the seller. For complex technologies, both the sales representative and the buyer are technically knowledgeable.

Resale can be especially profitable to both the facilities-based supplier and the reseller, even when the reseller's per-minute charge is less than the rates the supplier can charge its own retail customers. For example, Excel Communications successfully resells long-distance service in the United States. The company has minimized its marketing costs by utilizing multilevel marketing (MLM) to sell its service. Entrepreneurs representing Excel sell the service to their friends and acquaintances and seek out new sales representatives. These marketers are paid a percentage of their own sales and the continuing sales of those they recruit to sell more. Excel purchases at least some of its minutes from competitors such as AT&T and MCI, yet it manages to charge a lower per-minute price than these larger providers do.

Some corporations in other industries use representatives to resell their unused capacity. A corporation with an enormous long-distance volume can earn a per-minute rate significantly lower than the rate offered commercially to small users. Sales representatives sell minutes routed through the corporate network at a price that covers its costs and more, somewhere between the corporation's price and the listed price. The corporation can even benefit when the additional traffic raises volume and its corporate discount. The supplier benefits when new traffic is drawn from its competitors.

Resellers gain several advantages over facilities-based carriers. First, they enjoy the ability to enter virtually any market immediately. Resellers have the opportunity to achieve economies in advertising to large geographical areas and avoid the need to build facilities they are not certain they can sell. They do not need to borrow capital for investment or manage large networks. They can concentrate their resources on marketing and customer support, which are demonstrated to be decisive competitive factors. They can bundle services creatively and eliminate services from the portfolio with ease.

USN Communications sold its fiber-optic networks and switches in early 1996 and made a strategic decision to move to resale [6]. Management's reasoning was that the company needed to focus on its core competencies, customer service and billing.

One primary disadvantage of entering a market through resale is the loss of control over the product. For resellers serving the most price-sensitive customers, service quality is not as significant an issue. Other resellers compete face-to-face with the largest facilities-based carriers, and their customers demand comparable service. In theory, because resellers do often use the facilities of the giant carriers, the service is identical. If it is not, some resellers will opt to establish service-level agreements with their suppliers to ensure that their interests are satisfied. Service-level agreements are used by many corporate telecommunications users to guarantee the services of their facilities-based suppliers. The agreement can cover network reliability, speed of connections, or maintenance response times. Penalties for missing commitments could include payment credits or termination of the contract after a certain number of missed metrics. Service-level guarantees could provide an incentive for suppliers to bolster their networks to ensure greater reliability.

When only one facilities provider exists in a market the reseller wants to serve, resellers have less control over their costs and services than in markets with multiple wholesalers. The markets that do not attract facilities-based competitors are probably characterized by high costs and low profitability. For a reseller to succeed, it will need to keep its own controllable costs as low as possible.

Systems integration/value-added reseller

Telecommunications services are intimidating to nontechnical customers, and most customers are not technical. Many customers will need advanced technologies, such as voice and data networks, but do not maintain the internal expertise to support these technologies. Systems integrators and value-added resellers (VARs) provide a needed service, the ability to set up and maintain complex telecommunications systems.

Another layer of telecommunications infrastructure will emerge through software. Some software will be resident in the network itself, and configurable by the customer. Other software will reside in the customer's computer systems at the customer's site. Specialists will assist customers, either for a fee (systems integration) or bundled in the price of systems (VAR). Where the software resides and what it will do is still a matter of speculation. That telecommunications systems and computer software will converge further is the consensus of industry observers. Moreover, considering the growing reach of the Internet, and the growth of telecommunications and computing resources in all businesses, educational establishments, and the home, the specialist sector is certain to flourish.

Systems integrators will be sought by industry segments that for some reason decide not to maintain an internal telecommunications management division. Budget-conscious segments such as education, local government, and not-for-profit are often encouraged to outsource the installation, and some management, of internal telecommunications networks. For most mid-sized businesses and most industry markets, technology is not the primary mission-critical function. These companies do not want the daily management responsibilities for telecommunications systems, but their baseline requirements might be too complex for a simple retail solution. Industry-specific integrators and VARs will emerge to fulfill telecommunications planning and implementation requirements in niche markets.

Channel conflict

Channel conflict occurs when a company serves multiple channels and competes against its own customers. A wholesaler would ordinarily buy large volumes and resell them to retailers that sell the product or service to end users.

Suppose a wholesaler of telephone sets decided to use telemarketing to sell the product directly. In theory, since some costs are avoided, the wholesaler could sell the sets at a lower price than retailers, who need to add a large markup. In the end, this is generally counterproductive for the wholesaler. New costs are added, including selling expenses and customer care that is normally provided by the retailer. The retailers (the wholesaler's customers) would undoubtedly resent the added competition from their own supplier and could choose to represent a different wholesaler. The benefits accrued from the telemarketing venture would then be more than offset by the loss of customers or the loss of customer goodwill.

AT&T represents an excellent example of facing and then resolving channel conflict. At divestiture, AT&T retained both the long-distance business and the manufacturing business once called Western Electric. The manufacturing arm produced telephone end user equipment and network equipment such as large switches to support interexchange transmission.

AT&T's equipment brand was very strong, but its long-distance competitors were not willing to buy equipment from AT&T. In fact, even some of the local service companies were hesitant. A vertically integrated AT&T represented to them an important strategic competitor in local service and an actual competitor in local toll services. When other choices are available, companies do not want to give profits to their competitors, knowing that the profits will be used to compete against them in other businesses.

When AT&T's share of the long-distance market was 90%, losing a potential market of 10% was unimportant. The benefits of vertical integration accrued to its 90% share. When its market share fell to about 50%, AT&T realized that its value as two separate companies would be much greater if the channel conflict were eliminated. Consequently, AT&T divided its company into a telecommunications service provider and an independent manufacturing company, Lucent Technologies.

The idea that wholesalers/retailers have an apparent interest in the failure of their customers can be more damaging than their actual behavior. When a company's competitor is its supplier, all of its actions draw suspicion. Any other available alternative is appealing. This is not the corporate image that a wholesaler will desire in a competitive market.

As few companies are presently serving only the interexchange wholesale market, today's Internet backbone is provided by companies whose main business can span several channels. Regulators recognized potential service issues, so they required MCI to sell some of its backbone when it decided to merge with WorldCom. The market will help companies make these decisions in the future. An Internet service provider with a choice will connect to customers through a supplier that is not also a strategic competitor. Companies that concentrate on selling wholesale facilities will probably be the leading candidates to provide Internet backbone.

Early trends of local service providers

While no incumbent local exchange provider has announced an intention to split its operations permanently into wholesale and retail divisions, there is some indication that these providers are moving in that direction. Both Frontier Corporation and Southern New England Telephone (SNET) have restructured their operations into separate wholesale and retail divisions. It is worth noting—although not a complete surprise— that the companies first taking this initiative are among the least encumbered by federal regulatory scrutiny. WorldCom, a facilities-based competitive provider, supports both wholesale (WilTel Network Services) and retail operations.

At the other end of the spectrum, the RBOCs continue to debate with regulators concerning their resale and retail rates and their ultimate market positioning. Because they do not yet control their pricing and marketplace approach, they are understandably reluctant to form new subsidiaries without understanding their potential profitability.

Some of the incumbent carriers have created organizational divisions to support their wholesale operations. Ameritech's Information Industry Services division services network and information providers in 11 market segments. Bell Atlantic's Telecom Industry Services offers training, documentation, and an ongoing communications program. BellSouth's Network and Carrier Services division targets resale and interconnection customers at trade shows and other venues. These steps represent the carrier's recognition of the need to support wholesale markets without regulatory mandates. It is simply good business for a company to cater to intermediaries that offer to sell its products.

One benefit of separating the wholesale and retail business for companies that continue to serve both markets is that the company can conduct valuable analysis about the cost structure, profitability, and market readiness of each division. Eventually, should the parent company choose to sell one of the divisions, the attractiveness of the division is enhanced by both its organizational distinctness and its record of performance.

SELF-ASSESSMENT—CHANNEL STRATEGIES

The following questions will help telecommunications marketers to assess whether their channel strategies meet their goals.

- What channels does your company presently serve? Is your company in the desired tier of providers in each channel?

- What are your company's strengths? Which channels make the best use of these strengths?

- What are your company's weaknesses or competitors' strengths? Which channels are harmed by these?

- Do primary competitors focus on a single channel, or do they span several channels? What about emerging competitors?

- Has your company lost customers to channel conflict? What is the value of those losses? What potential customers would emerge if your company reduced the channels it served? What is the value of the potential gains?

References

[1] Bazzy, Jared, "Defense Companies Take Aim at the Telecom Market," *Telecommunications*, Americas Edition, Vol. 32, No. 9, pp. 40–46.

[2] Emmett, Arielle, "I Can Get it for You, Retail," *America's Network*, Vol. 101, No. 21, pp. 20–23.

[3] Salak, John, "Tangible Assets," *tele.com*, Vol. 2, No. 12, pp. 38, 40.

[4] Engebretson, Joan, "Resale Revisited," *Telephony*, Vol. 233, No. 11, p. 96.

[5] Finnie, Graham, "From Resale to Retail," *tele.com*, Vol. 2, No. 3, pp. 47–48.

[6] Lawyer, Gail, "The Troy Ploy," *tele.com*, Vol. 2, No. 6, pp. 64–66, 68–70, 74.

7

Direct Sales

Direct sales defined

Direct channels are the routes to the customer that are owned by the telecommunications provider. Employee sales staff, in-house telemarketing call centers, and customer representatives in company-owned stores are examples of direct sales.

While telecommunications service providers have always maintained sales forces, local service deregulation will substantially change the size, type, and requirements of the sales force and the overall sales process. Monopolists can have goals related to the size of orders or services upsold to customers, but their sales performance can only be measured hypothetically if there are no competitors. The expectation of imminent competition, coupled with the downsizing trends in the last decade, has placed telecommunications providers in a challenging situation. More sales activities will be necessary, but there are pressures not to increase staff. The cost of sales is rising while the prices of the services sold are falling.

When the marketplace is in a full state of deregulation, incumbent wireline carriers will undoubtedly suffer a loss in market share. The level of market share loss cannot be estimated, and it is almost pointless to decide what an appropriate or desirable drop in share should be. Wireline local carriers will undoubtedly track these losses conscientiously and compare their market share reductions to those of the other wireline companies. This is a useful but insufficient exercise. Other factors are also important to track: the quality of customers lost, changes in the volume of the remaining accounts, and the productivity of the sales force, for example.

The transition to deregulation

Deregulation will change virtually every aspect of the processes required for the sales of telecommunications services. Many of these changes are predictable because they are already in progress, or because they characterize similar deregulated industries, or because they characterize all other industries.

Qualified sales professionals will be in *high demand*. Until the market stabilizes, there will be far fewer sales professionals than there are positions to fill. In the transition period, the gap between the relatively few experienced professionals, and the requirement to hire a significant number of new sales representatives will probably result in larger salaries for the sales force. In the wireless industry, demand for sales professionals increased when PCS competitors entered the market, and the demands on the sales force have become more complex. This will occur in virtually every telecommunications market as they mature. Compensation will be higher than its present level in telecommunications sales to draw more labor. Moreover, sales compensation could exceed that of similar sales functions in other industries for an interim period until more sales professionals enter the industry market.

Deregulation will hasten the introduction of new technologies. The sales process will then become more *complex*, with new capabilities that the sales representative needs to understand. Second, as prices drop, customers with less technical knowledge than previous buyers emerge as the late majority. Accordingly, sales professionals need to bolster their technical expertise with communications skills to present the benefits of the product in a fuller description with fewer technical terms. The onslaught of new customers, coupled with lower prices, pressures providers to find effective sales channels with a lower cost or mix channels based on the demands of the selling process. Technological improvement also enables the development of vertical applications. The service can be specialized to a single market, adding to the complexity of the sale. The sales representative will need to identify the unique benefits of the service, while remaining conversant in the overall technology and retaining sales techniques and general interpersonal skills.

Monopoly telecommunications providers traditionally enjoyed a very small *turnover* compared with other industries. Moreover, the sales function in most industries has a higher turnover of employees than the companies as wholes. Therefore, in a deregulated marketplace, an increase in turnover in the sales function is inevitable, and it will call for both managerial and cultural changes in the telecommunications service provider.

There are many reasons supporting the low turnover rate in the telecommunications industry, and many, such as company loyalty, must be retained. Other potential reasons, however, could have a negative impact on future sales performance, including the following:

- Higher-than-market salaries;

- Low compensation risk;

- Reluctance to terminate nonperforming employees.

Sales management will need to conduct a critical analysis of the amount of tolerable turnover, the amount of existing turnover, and the present capabilities of the sales force. Sales management will also need to improve their recruiting and hiring practices to ensure that the sales representatives they want to retain are likely to stay and to perform well.

Higher turnover, coupled with a short-term lack of experienced sales personnel, is likely to enhance the value of a *college degree* among sales recruits. Monopolies are often characterized by employees, even senior management, who join the company immediately after high school and work their way to more responsible positions. Monopolies, aware that employees are loyal, are more willing than other companies to move employees through various departments and allow them to gain experience on the job. Sales personnel are often recruited within the company from operations or customer service functions. Companies that are more competitive need their employees to be productive very quickly and do not have the luxury of time for training. They prefer to hire sales representatives who have proven credentials or a college degree.

Besides the likelihood that seniority will become less important in the competitive telecommunications provider, *performance* will be much more visible and actionable. Poor performers will be reviewed frequently and fewer in-house transfers will be tolerated. Companies will be more selective when they encourage employee loyalty. Instead of setting goals to avoid turnover for all employees, they will target high performers and apply their resources to retaining those employees who contribute the most to the bottom line.

In sales, *measurement* will increase significantly, especially in the area of major account sales. Few companies in technology industries have kept salary-based compensation plans for the long term.

Selling will become harder until the market stabilizes. Much of the present customer base in deregulated markets such as long-distance and near-monopolies such as local exchange service requires little or no sales attention to continue to buy. Markets such as wireless, in which buyers are well aware of their alternatives, exhibit the largest amount of churn. Sales costs for wireless services are higher than costs for landline sales.

Companies will spend significant management attention on the percentage of *sales compensation at risk*, and the basis for that risk. On one hand, having too much of one's income at risk can create a sales force that is too hungry for sales or prevent new salespeople from earning a living wage during the required learning curve. The problem of the learning curve is sometimes addressed by providing a draw in the first year to salespeople who require experience or who serve large accounts where the sales time line is long. On the other hand, having too little of one's income at risk can saddle a company with the least productive sales force in the industry, with the most successful salespeople migrating to the provider with the highest compensation plan.

Changes *external* to telecommunications deregulation will cause the sales function to change as well. Retail concentration provides a venue for selling through additional channels, although selling telecommunications services through retail presents its own challenges. Well-funded overseas competitors entering the market places additional selling pressures on incumbent telecommunications providers and emerging entrepreneurs. The high-technology industry is characterized by labor shortages, which affect not only sources of technically qualified sales personnel but may also create backlogs in product development.

Needs of the buyer

The nature of the sales force will depend primarily on the needs and characteristics of the buyers. Buyers differ both in terms of their primary driver (service versus price) and the primary distribution channel (wholesale versus retail, with a hybrid role for a systems integrator) [1]. See Table 7.1.

In the wholesale segment, resellers will constitute most of the market that is driven primarily by price. Buyers in this segment are knowledgeable and have few barriers to switching. They seek commodity services and will probably choose to resell them without enhancement or with enhancements of their own development. The sales team to resellers

Table 7.1
Factors Driving Purchase Decisions: Selected Segments

	Wholesale	Systems Integrator	Retail Turnkey Services
Price-driven segments	Resellers	Education, local government, not-for-profit	Consumers, low-end
Service-driven segments	Multinationals, Fortune 500 companies, technology-dependent vertical markets	Mid-sized businesses and most vertical markets, in which technology is not the primary mission-critical function	Small business, SOHO market, telecommuters

needs to be knowledgeable about the technology as well as competitive offerings because the buyer will certainly have a similar knowledge level. Incentives to buyers will probably include loyalty programs and volume discounts.

Not all wholesale buyers base their purchase decisions on price. Large enterprises build networks for the use of their internal organizations. These buyers include multinational corporations, Fortune 500 companies, and technology-dependent vertical markets such as financial services, airlines, or call centers. The reliability of the network is essential to these customers, as a failure would shut their own businesses down for the duration of the outage. Buyers seeking exceptional reliability are not as price-sensitive if no competitor can provide both reliability and very low price. Like the other wholesale segment, the buyers are technically knowledgeable. They are more likely than the reseller segment to expect customized features from the wholesale provider. Because these large enterprises offer long-term sales and many decision-makers, the account team can deliver an effective and profitable sales approach.

Retail markets generally involve smaller volume sales and require more cost-effective selling solutions. The representative price-sensitive buyer in retail markets is the low-end consumer. The buyer is either knowledgeable but not loyal (such as a technology hobbyist), loyal but not knowledgeable (the customer who does not want to take the time to keep track of the multitude of offers), or neither.

The service-sensitive buyer in this segment is the small business owner. This buyer wants a low price but, like the consumer, cannot afford to take the time to learn about the market. In both retail situations, the telecommunications provider needs to find very low-cost sales approaches. Breaking the large, generalized segment into subsegments can lead to productive sales strategies. A provider that uses demographic studies to establish a bilingual sales force will be more effective than its competitors in neighborhoods that draw immigrants.

For the most part, long-distance providers have settled upon telemarketing as their cost-effective customer acquisition mechanism for both consumers and small business. They also utilize television and print advertising to encourage name recognition, but these venues do not close the sale. A follow-up call, either inbound or outbound, is required. For small business, the telecommunications provider can also advertise at trade shows or in trade publications. Other emerging and inexpensive alternatives include Internet sales, product tie-in programs, kiosks, and retailers.

The hybrid situation is the systems integration channel. Customers in this category have more volume than the retail category and require personal sales, but their volume does not merit the same level of attention as the wholesale category. Unlike the wholesale category, each customer does not constitute a major account. Examples of price-sensitive customers in this category are noncommercial enterprises such as education, local government, and not-for-profit organizations. Examples of customers that are less price-sensitive and more service-sensitive include medium-sized businesses and most vertical markets.

In the noncommercial, price-sensitive sector, contracts often keep the customer loyal for a limited time, but long-term loyalty is difficult to earn. The procurement process often in place in this sector is sometimes a barrier to the loyalty of any individual buyer in the organization. Management of national accounts or bid specialists can serve as a most effective sales approach in this market.

In the service-sensitive segment, buyers are moderately knowledgeable and moderately loyal. Unlike the wholesale segment, customers in this market sometimes rely on one technical staff member to make telecommunications purchases. Unlike retail, the buyer is both knowledgeable and potentially loyal. The purchase decision is often delegated to this staff member. Direct sales can be an effective approach in this segment.

Factors influencing the sales approach

The most effective sales approach depends on the *service to be sold*. Long-distance services are successfully sold through telemarketing partly because of their familiarity. Internet services have not been sold through telemarketing. Services with certain characteristics require sales that are more complicated:

- New services;

- Technical services;

- Services that are complex to use;

- Highly undifferentiated services with no significant price advantage.

More skills are required to sell products with these characteristics. Sometimes the provider strengthens the sales effort with promotions or infrastructure to ease the purchase of a service. For a new service, such as a custom calling service, the provider can offer a free trial period in the hope that the customer will learn to use the service and want to continue a subscription. For a technical service, the provider can offer kiosks that demonstrate the service in retail locations where the service is sold. Bell Atlantic's Knowledge Centers enable customers to see and touch products and consult with sales or technical staff [2].

The *length of the selling process* will also affect the resources to be applied to the sale. A multinational corporation in the market for a wide area network (WAN) will take months or years to make a purchase decision. A sale of this size and scope will require an account manager and perhaps a supporting team of presales technicians. This will ensure that the customer's questions and concerns are addressed promptly, and the sales team can provide a preview of the superior customer service that the prospect can expect after the sale.

Another factor influencing the sales approach is the *necessity that the service integrates* with other products and services the customer has already installed. The convergence of the telecommunications and information industries will ensure that integration will continue to be a source of challenge. A problem-solving sales force can ensure the success of a telecommunications provider by devising creative solutions that integrate existing infrastructure with new telecommunications services.

Last, the telecommunications provider that can offer a *single point of contact* to the customer will hold a significant advantage over those that cannot.

The sales process

The classical sales process comprises a variety of steps, including prospecting, lead qualification, the sales presentation, and closing. In a highly competitive market, especially a market characterized by churn, the sales process is constant.

The length of the sales process depends on the size and complexity of the sale. A single telemarketing call for local or long-distance services often contains the entire process through the close. If the telemarketing representative cannot close on the first call, the process is often ended.

Prospecting refers to finding potential customers. Virtually all homes and business establishments represent customers for basic telecommunications services. Ironically, one of the most common means for initial communication with consumers, no matter what the sales approach or channel, is the telephone. Many lists of prospects are available, in paper form, such as local telephone directories; on the Internet; or on a variety of computer media.

For telecommunications providers that plan to segment their markets or serve a market niche, list providers offer prospect information at a low price. Lists are available from professional brokers, and some lists can be found at libraries or on the Internet. Trade publications also provide lists of industry members.

Other sources of prospects include referrals, networking, and cross-selling with other providers. Inquiries from potential customers also qualify as prospects.

Sales leads always require *qualification*. The qualification process acquires more information about the prospect, calculates the likelihood that the customer will buy, determines if the customer can pay for the product, and evaluates whether the customer is desirable enough to pursue further.

Inadequate qualification can result in lost sales because customers do not have buying authority or budget, uncollectible bills because creditworthiness was overlooked, or sales efforts wasted on customers who are not ready to buy.

This process is relatively new to providers with a history of monopoly service and a culture of serving all users. Providers qualify leads as a means to apply scarce sales resources to the most productive opportunities.

The *sales presentation* is the opportunity to present the service and its benefits to the prospective customer. This phase can take a minute or several years. Within the presentation, the sales representative is expected to provide information to the customer and overcome any objections that the customer raises.

Closing is one of the most difficult selling tasks. Whether the sale is a brief telemarketing encounter or a lengthy procurement process, the sales representative needs to recognize the time to close the sale and get the closing accomplished. Closing involves recognizing buying signals, handling any remaining objections, stating clearly the nature of the sale and its associated benefits, and finalizing the sale.

Sales to new customers reportedly cost about five times more than sales to existing customers. Cross-selling and up-selling services to the existing base can be productive in volume and profitable for the telecommunications provider that makes the effort. Frontier Communications found several techniques to improve its sales results [3]:

- Setting monthly expectations for its customer service employees;

- Posting results for all employees to see;

- Paying commissions to employees for their sales results;

- Employing coaches to train groups of employees;

- Taping sales calls for critique by employees and their coaches.

Because of this program, some markets that were previously contributing 5% of the carrier's total sales grew to about 30% of total sales.

Comparative channel sales costs

An important part of the sales channel decision is the cost of the channel and its anticipated effectiveness. According to *Sales and Marketing* magazine, the average sales call costs $106.21 in the service industry [4]. Sales that take several or many sales calls to close must provide enough profit to justify their effort. While enterprise sales of large telecommunications networks certainly require a long and personal sales process, most telecommunications sales transactions are too numerous and too small for the direct sales channel.

In the telecommunications industry, sales costs per order vary by channel from about $30 per new customer to more than $400 for a sale brought in by an agent [5]. The typical RBOC inbound call center can acquire a new customer for $40–70. It should be noted that, at present, the typical residential customer is not aware of, or interested in, competitive local service options. While telemarketing is an effective and efficient sales channel, it would be naive to expect the same level of customer inertia in the future as has historically occurred.

For wireless, acquisition costs are high, compared to landline customer acquisition. This discrepancy could be due to differences in the level of market competition or to the higher cost sales channels for wireless sales, but the reason is less important than the trend it probably signifies. Providers of competitive wireline and wireless services should expect higher channel costs when telecommunications markets become more competitive.

Nevertheless, the cost of a channel in itself is not a sufficient base for a sales strategy. For a large, long-term sale, buyers require a direct sales representative. A high-volume sale will easily cover even a large acquisition cost of the customer, and, of course, no other channel can provide by itself information that the buyer needs to make the purchase decision.

Telecommunications providers will utilize a variety of sales channels, measuring both the cost and effectiveness of each to create the optimal mix of selling resources and cost. Testing sales channels will include moving sales efforts from higher-cost channels to lower-cost channels and evaluating sales results to determine whether the change should hold. In addition, telecommunications service providers will be vigilant in eliminating unnecessary costs in all sales channels to increase profitability.

Sales effectiveness

According to strategy consultants Booz-Allen & Hamilton, company growth is strikingly tied to the ranking of the sales force. A superior sales force could affect the company's revenue growth by as much as 40–50%.

Sales force automation (SFA) is gaining in popularity for increasing the effectiveness of the sales function. SFA can lead to improvements in operating efficiency, sales productivity, and measurement. While many SFA projects have encountered barriers in the organizational culture or failed for a variety of reasons, success stories occur. As the technology improves and companies gain more experience in establishing these systems, sales force automation is expected to grow.

One barrier to sales effectiveness is the traditional but inaccurate view that sales behavior cannot be measured. New approaches to sales management, accompanied by information technology, are enabling companies to isolate the variables that are identified with successful sales. One approach is activity-based costing, which provides a portrait of the activities conducted by the sales force, their proportion of time, their concentration among employees, and their impact on sales success.

Developing and using objective measures to assess the effectiveness of sales requires the following steps:

- Defining measures that match the objectives of the enterprise;

- Defining processes to obtain accurate measurements;

- Developing a compensation structure appropriate to sales risks;

- Rewarding excellent performers and eliminating nonperformers;

- Revising measures for improvement or to meet new market requirements.

The process begins by defining measures that match corporate objectives. The sales effectiveness process requires a commitment to measurement, outside of the requirements of financial statements and compensation plan components. These measures need to have sufficient detail without unnecessary complexity. The dependent variables for measurement include the following:

- Growth indicators such as new sales or market penetration;

- Profitability indicators;

- Ratios that demonstrate the relationships between revenues and their associated expenses;

- Measures that recognize the activities and priorities of the most effective sales professionals.

If possible, measures for customer service can be included, if they are within the control of the sales organization.

Defining processes to obtain accurate measurements is essential before sales activity is initiated. These processes will ensure that the necessary data are collected during sales activities and that follow-up data are matched to them after sales are completed. It is often tempting to use the documentation created by other corporate systems as the source of sales measurement, and some of the time these data are adequate. When it does not capture the exact measurements needed, new systems should be devised, even if their only use is to evaluate sales performance. Outside service bureaus may have more sophisticated sales measurement processes than can be developed or installed in-house. Quality of measurement processes can be a selection criterion between several outsourcing candidates, and the existence of excellent measurement systems can constitute the justification to outsource certain sales activities.

Once the objectives and the measurement systems are known, compensation structures can be established.

Compensation structures

Sales compensation and motivation are among the most important strategic decisions for a competitive company to make. The aggressiveness of compensation plans will affect the composition of the sales force and the level and type of sales. Designing compensation plans must be tied inexorably to corporate goals and strategies and accompany the highest-level business planning efforts.

In many companies in highly competitive industries, it is not uncommon for the top sales performers to earn more in a successful year than most of their management. The transition to deregulation has placed upward pressure on the compensation packages awarded to executives [6], and rewards are increasingly tied to performance. While this is a normal occurrence in industries undergoing deregulation, this can have the long-term effect of raising sales salaries for those sales professionals serving the most important accounts.

Often, sales of low-volume services are sold through inexpensive channels such as direct marketing or commission-only arrangements. One possible drawback to commission-only compensation is that sales representatives can become overzealous in their closing efforts. The proliferation of slamming complaints is potentially an indication that this is taking place in some distribution channels.

Slamming is the practice of signing new subscribers to long-distance (or any other) service without the explicit consent, or sometimes the knowledge, of those subscribers. Subscribers do not realize that they have been slammed until a bill arrives from an unknown or unexpected telecommunications carrier. A more benevolent explanation for the high incidence of slamming reported in the United States is that customers are confused by the changes in the industry and do not realize that responding to a telemarketing call results in changing their carriers. Slamming results in high fines to the carrier involved, so several long-distance carriers have instituted audits on telemarketing calls. Customers are sometimes required to verify the change of provider to a supervisor, repeat the transaction on audiotape, or sign a "contract," which could be in the form of a check.

Some slamming and other self-defeating selling tactics occur because most compensation plans are designed to increase sales, not profits. They almost never contain provisions for long-term customer retention, nor do they require profitability on sales. In general, sales professionals have enough challenges without being responsible for customer longevity or profit. However, comprehensive compensation plans ensure that the sales force will apply their efforts toward those sales that indeed add to profits and customer loyalty. Those are the objectives of the compensation plan designers, not the sales representative.

The purpose of a sales compensation plan is to encourage the sales force to sell services in a desired mix and at a level that meets corporate objectives. It is fair to assume that an efficient sales force will not actively sell services that are not contained in their own compensation plan. Similarly, capable sales representatives will sell services that are easiest to sell first, unless the compensation plan accounts for the difficulty of one sale over another and evens out the reward.

For this reason, many compensation plans provide higher commissions or bonuses to salespeople who sell to new accounts, rather than a similar sale to an existing account. Companies often create bonus incentives to sell new products to any customer.

Companies use incentives to bolster corporate strategies. For example, analysis of a particular service line might demonstrate that the cost of supporting its existing customer base will soon exceed its revenue stream. It would be important in that situation to migrate customers to a newer, compatible offering without losing the customer to competitors by eliminating the service line. A special bonus can be offered for sales that move the customer to the new line. The corporate strategy can also include an attractive upgrade price. A lower-than-normal upgrade price might discourage the sales representative from embracing corporate goals unless the commission structure made allowances for the extra effort.

Sales quotas strive to be objective, but they must be tailored to the requirements of the environment and the capabilities of the sales representative. New sales representatives usually start with a lighter quota and a time frame designed to bring them to a higher level. Sales representatives with the best territories should have the most challenging quotas.

Managing the sales function

The responsibility of the sales manager is to help sales representatives meet their stated goals. Effective compensation plans are structured so that the sales managers have a significant stake in the success of their subordinates. Sales managers also contribute to the compensation planning process. Because of their proximity to the sales force and the customer, sales managers can provide valuable input to compensation plans. They can help to incorporate value-added goals that do not require additional revenues, such as customer satisfaction, account management, cross-sales of services outside of the quota mix, team efforts, and long-range goals, without immediate revenue or commission impact.

Other activities that can improve sales performance include collaborating with representatives on their goals, personal recognition by management and peers, and contests to create a constant level of activity when sales quotas are annually measured or too long-term to maintain a high energy level among the sales force. Another valuable contribution by sales management is to meet with managers in all corporate and operational functions to emphasize their own roles in customer satisfaction and the sales process.

While rewards to sales professionals frequently take the form of higher commissions or bonus payments, other nonmonetary incentives can be effective in encouraging superior sales performance. Successful representatives are rewarded with expanded territories. Conversely, companies that divide the territories of the most successful sales representatives run the risk of losing their high performers.

Sales representatives require coaching and feedback on their performance, but by nature, individuals drawn to sales are self-motivated. One of the most valuable contributions that sales managers can make to their subordinates is to negotiate with their own management to do whatever is required to close a sale. Sales managers are responsible for keeping sales costs to a minimum by balancing the need for sales expenditures against unnecessary costs. They can also reduce the cost of sales by ensuring that sales professionals are not required to undertake inessential administrative tasks.

Sales management always requires the distasteful task of firing nonperformers. In a competitive market, the cost of retaining nonperformers is a combination of the expense of the sales representative, the work performed by others in support of lost accounts, and the opportunity cost of accounts not sold.

Sales personnel require training in both sales techniques and product knowledge. Some customers will require sales representatives to hold a broad understanding of the technology under discussion and the customer's own industry concerns. For the most complex sales, the account representative can reinforce his or her own expertise with technical presales specialists who are most knowledgeable about the technology. Beyond product and industry training, general sales training, such as training in the sales process, in developing creative solutions, and especially in negotiations is critical. Successful selling is not an art; it is a skill.

SELF-ASSESSMENT—DIRECT SALES

Telecommunications providers need to know the answers to questions such as the following.

- What is your sales strategy? Has it changed as the market has become more competitive?

- How do you evaluate the mix of sales channels? Do you measure each channel on its effectiveness and its cost?

- Are your sales compensation plans designed to attract the sales representatives you want to hire?

- Do you support multiple sales forces? Are they structured to meet the needs of your company's organization or the customer?

- Are your various sales forces staffed adequately? Are they properly trained?

- What is your sales effectiveness? What measures have you taken to improve the effectiveness of your sales force? Have training and development programs been launched or modified? Have recruiting processes been modified as required?

References

[1] Strouse, Karen, "When the Dust Settles, Will LECs and IXCs be Segmented? Or Decimated?," *Telephony*, Vol. 235, No. 10, pp. 27–32.

[2] Wood, Wally, "Corporate Anorexia: Can Telcos Survive?" *Telephony*, Vol. 230, No. 10, p. 22.

[3] Pappalardo, Denise, "The Frontier Way: Turning Customer Calls into Sales Calls," *Telephony*, Vol. 231, No. 17.

[4] Marchetti, Michele, "Hey Buddy, Can You Spare $113.25?," *Sales and Marketing Management*, Vol. 149, No. 8, pp. 69–77.

[5] Docters, Robert G., "The New Wholesalers?," *Telephony*, Vol. 234, No. 4, pp. 26–34.

[6] Driscoll, Kathleen, "Executive Compensation," *tele.com*, Vol. 2, No. 13, pp. 81–83.

8

Indirect Sales

The indirect sales channel

Among the many channel management decisions telecommunications providers will need to make is whether to use external—or indirect—sales. *Indirect sales* are sales made through channels that are contracted, not employed, by the telecommunications provider. The most common form of indirect sales involves agents selling on behalf of the telecommunications provider. Indirect sales provide additional marketplace leverage for telecommunications providers by creating new sales with a minimum of incremental expenditure.

Industries characterized by significant growth or churn are excellent candidates for indirect sales, which limit the provider's fixed expenses and commitment to full-time employees. Providers can enter and exit markets much more freely when they take advantage of indirect sales channels. When the local telecommunications services market becomes as competitive as the long-distance market in the United States, the agent channel is likely to increase. While some of the companies presently operating as local resellers of telecommunications services will undoubtedly become facilities-based carriers on their own, others will act as agents for the facilities-based providers.

Examples of agency abound: sales agents, alliances, affinity groups, telemarketing service bureaus, independent sales representatives, distributors, retailers, and direct marketers that sell the product through catalogs and mail. What is common among these sellers is that they sell the telecommunications services in the form of its original brand. Thus, a consortium of large international providers do not qualify as indirect sellers when they create a new brand for the services sold under the umbrella of the consortium. Affinity groups such as clubs or organizations serve as agents when they sell the telecommunications service within its original brand name. Affinity groups tend to retain the well-known branding of the services they sell to increase name recognition and reassurance among their customer base. In 1998, GTE engaged a marketing firm to create affinity programs to reach residential long-distance customers.

Companies that rebrand a telecommunications service are not agents; they are resellers. Resellers are generally more entrenched than agents are; they sign a contract with a telecommunications provider to purchase a certain number of minutes and they are responsible for other customer service functions. Beyond the brand name, the biggest difference between agents and resellers is that the reseller owns the customer. When an agent sells the services of a telecommunications provider, the provider, not the agent, retains ownership of the account. Resellers themselves are among the most active users of agents to sell their services.

Agents in the telecommunications market

Telecommunications providers use agents for virtually every service currently sold to customers. Agents represent providers for long-distance services, cellular service, paging, callback, and local service. Agents often serve those customers who prefer face-to-face contact with the sales representative but do not purchase sufficient volume to interest the largest carriers' direct sales forces. Agents often operate in the small-to-medium sized market for business customers or the high-volume residential customer. Customers below these thresholds are served with lower-cost channels such as direct marketing through telemarketing and bill inserts.

According to strategy consulting firm ATLANTIC-ACM, independent agents will account for about one-quarter of the U.S. interexchange service revenue outside of AT&T, MCI, and Sprint [1]. Resellers' use of agents increased 11% from 1995 to 1996, and switchless resellers draw nearly a third of their business revenue from agent sales [2].

The largest carriers use indirect sales to boost their revenues. Through its comprehensive program targeted at VARs of computer technologies, GTE qualifies sales agents to sell an array of services including voice, data, video, wireline, and wireless. GTE utilizes agents to sell pre-paid cards redeemable for its land-based long-distance service and its proprietary Airfone service while traveling by air.

Sprint supports a program that authorizes agents to sell Sprint-branded products and services. Its entry qualifications for agents include a requirement that the partner already supports an existing customer base that is complementary to Sprint services, holds a business plan that targets customer retention, and has more than one year's experience. MCI WorldCom uses agents to sell highly technical offerings such as frame relay services. The company requires revenue commitments, training, a management relationship, and a credit background check, among other criteria. Local exchange carrier BellSouth has used agents since 1985 and increased its agent ranks by 30% in 1997 [3]. US West sponsors a program in which qualified agents sell branded US West network products and services.

According to Virginia-based *Cable and Wireless,* the monthly revenue from agents rose from about 5% to 25–30% of total revenue in a period of five years [4]. The agent channel is expected to grow, aided by the new services that will spring from a deregulated market and the entry of providers such as ISPs and CLECs. These companies, most often smaller than the incumbent carriers, are the best positioned to use agents. New entrants to the competitive market also target those customer segments that are often served by agents, the small-to-medium sized business customer.

Mid-sized and emerging facilities-based local service providers such as TCG, which is now owned by AT&T; Tampa-based Intermedia

Communications, Inc. (ICI); and Maryland-based e.spire use agents to sell in their targeted business markets. Agents offer a cost-effective sales force for facilities-based providers entering new markets. Building the local facilities to serve new geographical markets requires significant capital investment, so using agents for sales helps to reduce the initial cost of entry and the associated business risk.

Wireless companies have used authorized agents since cellular service began and have extended the concept to include a new collection of authorized agents, selling their branded services, within retail stores.

Callback providers selling discounted international calling use agents extensively, partly because their customer base is geographically dispersed. According to market researcher M.J. Scheele & Associates, 72% of callback sales stem from independent agents [5].

Master agencies represent multiple agents. Using master agencies rather than engaging individual agents has several advantages to the telecommunications provider and to the agent. The master agency often guarantees a certain level of calling volume in exchange for discounted rates to sell. This arrangement resembles the reselling arrangement, but the master agency does not rebrand the service. The agency frequently provides other services on behalf of the telecommunications provider. These services—such as hiring, handling the customer provisioning, offering customer care, and training—can upgrade the quality of the agent base and reduce the cost of using agents. Meanwhile, working with a master agency enables the agent to delegate many of the administrative support activities that an independent agent would have to carry out itself. The agent is then free to spend time selling and increasing revenues instead of performing non-revenue-producing functions.

Excel Communications used multilevel marketing successfully in the residential market and grew to one of the top resellers in the United States. Its success is especially noteworthy because the residential market is very low-margin and probably cannot support a traditional agent channel. Its marketing structure enabled its agents to receive commissions for the services they sold and discount long-distance service as well. The pyramid structure of new sales engenders loyalty among its agents.

In some markets, agents are expected to sell one provider's product exclusively. In other telecommunications markets, agents can sell from a variety of providers. This enables the agent to choose from an array of offerings to customize the solution to the needs of the customer. In addition, the agent can bundle products from several providers when each provider alone does not offer the bundle of services desired by the customer.

The cost-benefit analysis

Agents are most valuable when the following conditions exist:

- When a telecommunications provider needs to penetrate a new market with a relatively small investment;

- When the provider serves the small-to-medium sized market;

- When the provider would like to serve a niche that requires technical product knowledge that its internal sales force does not presently have;

- When a niche market requires industry rather than technical knowledge.

Telecommunications providers can utilize indirect sales channels for all of their sales or none of their sales or use some in-house sales resources and some outside agents. A large part of the decision is financial. Agents generally do not receive compensation other than their commissions. They are not reimbursed for their selling expenses. Therefore, an agent who sells nothing costs nothing.

In-house sales representatives can receive only commission, or a small salary and commission. Additional costs of an in-house staff include office space and supplies, out-of-pocket expenses, and supervisory management. The commissions for the in-house sales force are typically lower than the commissions for outside agents. There is a break-even point for in-house versus outside sales because the lower commission structure for the in-house representatives is more profitable at the highest level of sales. Similarly, a huge sales force of agents will require in-house management and other infrastructure, eliminating some of the cost advantage. This is a straightforward calculation, once the cost factors and revenue levels are known. Other issues beyond cost should affect the mix of direct and indirect sales resources.

In general, providers entering new markets will benefit considerably from using agents and no internal resources until they reach a certain critical mass of revenues. A provider expanding into new markets will always benefit from using agents, even if in-house sales are used in more mature markets. Other factors can justify the long-term use of agents.

Benefits to the provider

Using indirect sales channels offers several benefits to the telecommunications provider. Most of these benefits increase market share or reduce the provider's administrative costs.

Indirect channels provide better *coverage* to enhance the efforts of a direct sales force. Agents and other indirect sales channels enable telecommunications providers to enter new geographical markets without adding significant administrative infrastructure for marketing and sales and their associated management. Indirect channels selling a startup company's brand in new markets enable a company to appear bigger than it is. For small companies, agents provide a sales organization in any area that the company desires to serve. Agents can represent the only option for a new company that wants to invest its startup capital in facilities, not salaries.

In near-commodity markets such as long-distance, the agent channel offers *low-cost distribution*. If the local service market becomes as intensely competitive as the long-distance market, it will make sense for the new providers and the incumbents to use agents for sales to suitable customer segments in the local market as well. Telecommunications providers already using agents can simply expand the service lines sold by their agents.

The low legacy price structure of local service discourages agency sales during the transition to deregulation. If a service is already sold at or below cost, few providers are eager to increase its sales. Agents who sell local service under agreement with a local service provider often require the service package to include long-distance service, which boasts better margins. When prices stabilize to a cost basis as the market matures, local service should become profitable and desirable to agents and providers.

Unlike an in-house staff, agents do not receive salaries, and they incur *no fixed costs*. The structure of most agency agreements is based on commissions on services sold. Therefore, agents only cost money when they bring in customers. Most often, the level of commission is tied to gross sales, so the agents who earn the most are the ones whose customers provided the most revenue. Because agents are independent, they incur no administrative costs such as office expenses or payroll taxes. Even telemarketing agencies, with their large infrastructures, are often compensated on a commission basis.

Some agents earn a bonus for new customers but receive no recurring revenues. This arrangement still offers some of the benefit of commissions in that the agent is not paid until the customer is signed. Again, fixed costs are eliminated. In other arrangements, the agent receives a bonus at the initial customer signing and smaller residuals for the duration of the customer relationship. This encourages the agent to oversee the customer account and work to eliminate churn.

Marketing skills constitute another benefit of using agents, especially for the leading telecommunications providers. The job of an agent is to sell, and the typical agent is drawn to sales and experienced in the skills needed. Incumbent telecommunications providers can outsource their sales to agents while they develop a competitive internal sales force. Alternatively, they could decide that in-house marketing skills are not a core competency, and continue to use indirect sales while developing more strategic skills in-house.

In markets served by agents that do not represent a product or an industry niche, the agent does not require a great deal of technical knowledge. In the rare case that technical expertise is required to complete a sale involving an agent, the telecommunications provider can send a technician. With agents, telecommunications providers can thus bolster their marketing skills where they are weak and provide other sales support in areas of strength. Telecommunications providers can also choose to use agents to sell lower-margin, less technical services such as voice and use trained technical sales representatives to sell sophisticated data services.

Agents frequently require *less supervision* than internal employees, and are used to operating independently. This reduces the overall cost of indirect channels as compared to direct channels.

For smaller telecommunications providers or large providers entering new geographical markets, agents often bring *local knowledge* and leads to the relationship. Agents eliminate some of the risk in serving new markets. A facilities-based provider will undoubtedly need to make a significant investment in facilities when entering new markets, but there is less need for brick-and-mortar offices or management to supervise the new staff. Furthermore, local agents are likely to have a customer base from their previous sales experience from which to draw new customers.

Agents who sell network capacity are selling *incremental network utilization*. For a highly cost-intensive industry such as telecommunications services, the incremental customer delivers the same revenue at a much lower cost than the earlier customers do. Agents provide sales at a cost-effective rate and can help to achieve critical mass in new markets, or boost network utilization in faltering markets.

Benefits to the agent

Agents gain from the relationship several benefits that are not available in other sales alternatives. They enjoy more freedom from daily supervision than the provider's internal sales force, and they incur much less business risk than resellers or facilities-based providers. Because they never take ownership of any hardware or software in their portfolio, they make no up-front investment. Agents can decide how much they would like to work, as long as they are willing to accept the lower commissions arising from fewer sales when they work on a part-time basis.

Independent agents have the advantage of selling a branded product to customers at very competitive rates. They can obtain sales experience and training from the telecommunications provider, making them more valuable to a future employer or preparing them to start their own telecommunications services business. If the agency agreement pays commissions for the life cycle of the customer account, the agent earns residuals on business sold in the past, as long as the agent can prevent the customer from leaving. The agent that generates a large customer base can enjoy large monthly commissions.

Master agencies enjoy some of the economies of scale that resellers achieve, while maintaining their independence and avoiding the overhead and regulatory requirements of resellers. They can represent a variety of carriers, bundle services in ways that may be unavailable to the providers they serve, and control the relationship with the customer.

Generally, agents are protected from many of the obstacles facing the providers they represent. They are not necessarily responsible for providing the service or the customer support after the customer has subscribed. Some agents do prefer to control the after-the-sale relationship with the customer and choose not to use the provider's customer care. Most agents are not penalized for uncollectible customer bills.

Potential drawbacks for the provider

The agent-service provider relationship can be strained when the objectives of one party run counter to those of the other. All contractual relationships have the potential for conflict, and the agency relationship is no exception. Furthermore, the emergence of agents was initially mandated by regulators, and incumbent carriers were inclined to respond in a lukewarm manner. Agents proved to be cost-effective and productive, so the relationships continue, but cultures sometimes take a long time to transform.

The telecommunications provider that uses an agent has *less control* over the agent's activities—the counterpart of not needing to manage them. Supervisors who cannot restrain themselves from micromanaging their agents will experience attrition, especially among the high performers.

The agency agreement clearly states that the customers legally belong to the telecommunications provider. Nonetheless, the agent is the person that the customer knows, the agent is the problem solver, and the agent is the customer's point of contact. As long as the agent is content in the contract with the telecommunications provider, the customer relationship is safe. In an agency relationship, the telecommunications provider entrusts the agent with the customer's day-to-day concerns. While the agent will certainly provide some feedback when customers have problems that require the provider's attention, the agent will quietly solve the less critical issues. This is a benefit of using agents, but it can result in a *scarcity of customer data* that could benefit other sales agents or management.

Some agents are responsible for their own *training and development*. Telemarketing firms often differentiate their services from those of their competitors by their superior training programs. Other companies that represent multiple agents can provide support to their agents in the form of sales training, technology sales tools, and collateral marketing materials. Individual agents can provide their own training or look to the telecommunications service provider they represent for training and development. While either alternative is acceptable, each has its drawbacks. If the telecommunications provider offers the training of its agents, it loses some of the cost advantage of using indirect channels in the first place. If the telecommunications provider depends on the agent to provide its own training, it loses the control it would have when all the training is conducted in-house. Without identical training for agents and the internal sales force, the two channels will act differently. Comparisons between the channels could be biased because of the differences in training. This issue is not as important for sales of services that are not highly technical, or for agents who are already experienced, or when services are near-commodities and require less product knowledge.

Agents frequently represent a *variety of providers*. Some agents are not willing to sign agreements that prevent them from selling the services of other providers. Some telecommunications providers are satisfied that the agent will sell the right product to the right customer, and others mandate exclusivity. The advantage of having an agent sell an array of products is that customers appreciate the one-stop shopping that continues to elude most providers that sell only their own service offerings. When a provider's service is competitive and priced aggressively, sales are not harmed when multiple vendors are available. Nevertheless, when your competitor's product is the most attractive to the customer, but the competitor's compensation plan is more rewarding to the agent, the agent can neglect to offer your superior service to the potential customer. Thus, this conflict, if it occurs, benefits neither you nor the customer.

Indirect sales channels sometimes take control of operational functions, including setting up the service and customer care. *Customer care* is a vital function, and many providers are reluctant to give up control. Moreover, the telecommunications provider undoubtedly supports its own direct sales customers with a customer care infrastructure. The agent's customers do not represent a large incremental cost. If customer care is critical to the telecommunications provider, this area can be a part of the original negotiations with the agent. This issue is the reverse of the benefit of reducing costs that agents provide.

Some telecommunications service providers are concerned that ineffective agents, operating independently, can be a detriment to the sales force. Agents counter this argument by asserting that agents who lose customers will eventually exit the business for a lack of earnings. While this is true in the long term, the agent, like the internal sales force, carries the *reputation of the service provider*. Agents, by their lack of supervision, could do some damage to a company's reputation simply because the signs are not as apparent to management as they would be with an in-house sales representative.

Similarly, some telecommunications providers fear that a commission-only fee structure can result in the agent signing a large number, but a lower quality, of customers. *Slamming*, the practice of signing customers who are not aware or do not authorize the sale, can result. Sales representatives, in their zeal to earn commissions, could sign customers without proper lead qualification. This is not a concern that is particular to using agents; it would affect in-house sales representatives in the same way if their commission structure were as aggressive. A telecommunications provider depends on sales, direct and indirect, and needs to provide adequate screening and other procedures to ensure that unwanted customers are not subscribed.

Unfortunately for the telecommunications service provider, the most successful agents can be the most *likely to leave*. Acting as an agent provides the most flexibility compared to resale or facilities-based startup, but it also offers the lowest financial reward to the sales agent with high-volume sales. The agent can earn more by becoming a full-fledged reseller of the telecommunications provider's services. For those agents willing to take additional risks, becoming a facilities-based provider is even more attractive.

According to a recent ruling, CLECs are permitted to target customers they formerly served as agents [6]. A successful agent of local services, CTC Communications Corporation decided to become a CLEC rather than an agent of Bell Atlantic, which uses more than one hundred agents. The ruling followed a protracted dispute between the two former partners in sales. Another Bell Atlantic agent, Net2000 Communications, has become a CLEC under more favorable circumstances [7]. Both of these agents were among Bell Atlantic's top performers before they left.

Frontier Corporation learned the hard way that depending on outside sales can significantly affect the level of business. Frontier acted as a wholesaler to Excel Communications, in a relationship that resembled an agent relationship, with Frontier selling at both the wholesale and retail levels. In 1996, Excel was responsible for nearly a quarter of Frontier's long-distance revenues; in 1997, Excel represented only about 6%, partially causing Frontier's 17% drop in long-distance revenues [8].

Potential drawbacks for the agent

Being an agent rather than a reseller or a facilities-based provider can have drawbacks and risks compared to the agent's alternatives. Agents, being such a small part of the telecommunications provider's overall sales, have little control over the service portfolio they sell. Some agents are frustrated by the regulations and tariffs imposed on the companies they represent, during the transition period in which rules are not evenly applied to all providers.

Also, not all agents require the flexibility of the arrangement and would prefer the higher earning potential of becoming a reseller or a facilities-based provider. The agent experience, for these individuals, is a valuable, but temporary step on the road to becoming an entrepreneur.

When agents are in exclusive sales relationships with a single provider, they are captive to the telecommunications providers they serve. The abundance of mergers in the telecommunications industry can threaten the future of an agency agreement with a carrier. If the telecommunications provider is very large, and the agent is independent or in a small master agency, it is difficult for the agent to wield power in the contract negotiations with the carrier. Agents, in an attempt to respond to their own customers' needs, can become frustrated with the responsiveness of large companies that are setting priorities to address the diverse needs of an enormous customer base.

As the agent develops a large customer base and if the agent can find funding sources, the migration path for an agent is to resale and then to facilities-based service. Not all agents will be attracted by these alternatives, but those who are will eventually leave the flexibility and protection of agency relationships.

Managing agents

In some ways, hiring and managing agents is similar to managing an internal sales force. Sales management involves vigilance in measuring sales productivity, customer satisfaction, churn, and other indicators of overall sales performance. The sales manager must also monitor direct and indirect channels to ensure that the mix of in-house sales representatives and agents is still appropriate to the market.

Selecting and supervising agents requires less attention day-to-day than similar functions involving internal sales representatives. Nevertheless, the agent relationship is a partnership and is a substantial responsibility. The high intensity of competition in most telecommunications services markets puts enormous downward pressure on prices and on their underlying costs. This places stress on the agent relationship in two ways. First, the provider needs to reduce costs to the minimum, and agent commissions are a cost of doing business. Second, in the long-distance and wireless markets, the markets have grown, but the prices of services have dropped considerably in a few years. Agents who earn a percentage of gross sales will have to increase their sales levels just to earn the same amount over time.

Telecommunications providers need to offer agents the same information support they provide to their internal sales representatives. Agents need to have access to customer accounts, the status of provisioning, billing statements, and any other information that helps the agent retain the customer. Similarly, agents need to provide an adequate amount of information about their accounts and their sales activities to management, so issues can be resolved promptly.

Contracts between agents and providers need to be explicit about compensation, the accountability of both parties, and the conditions of termination of the agreement. Agents will expect the telecommunications provider to live up to its commitments for provisioning or customer care. Providers need to define not only the rate of commission but also the timing of payments to the agent and provisions for bad debt, fraud, or slamming as applicable. Terminating the relationship can become complicated. Though the provider owns the customer account, the agent often controls the customer relationship. When the contract is over, because the provider exits the business or merges with another carrier, or because the term expires, contracts must be clear as to the actions of both parties concerning the customer accounts.

SELF-ASSESSMENT—INDIRECT SALES

Several questions will help telecommunications marketers to determine whether indirect sales are warranted.

- Is there a process in place to compare direct and indirect channels?

- Are indirect channels more or less expensive than your current sales channels?

- How effective are the internal and existing external (indirect) channels in achieving sales goals?

- Is it appropriate to use indirect sales for new geographical markets?

- Is it appropriate to use indirect sales channels for new product introductions?

- Are some of your existing markets candidates for agents?

- How do your competitors use agents? Is their use of agents helping their competitive positioning against your sales structure?

- What are the implications on network costs of increased utilization? Can you use indirect sales channels to increase usage of the network?

References

[1]   Titsch, Bob, Jr., and Peter Meade, "Diving for Dollars," *Phone+*, Vol. 12, Issue 12, pp. 40–44, 121.

[2]   Reed Smith, Judy, "Channel Over Churn," *Phone+*, Vol. 11, Issue 8, pp. 56–57, 68.

[3]   Lynch, Karen, "Unsecret Agents," *tele.com*, Vol. 2, No. 14, pp. 38, 40.

[4]   Kim, Gary, "Agent Channel: A No-Brainer," *Phone+*, Vol. 11, Issue 15, pp. 48–51.

[5] Warren, Robin, and Cathleen Woodall, "Far-Flung and Flourishing," *Telephony*, Vol. 232, No. 21, pp. 26–34.

[6] Carlton, Michele, "CTC Settles Dispute with Bell Atlantic," *Telephony*, Vol. 235, No. 8, p. 64.

[7] Lawyer, Gail, "Breaking Ranks," *Phone+*, Vol. 12, Issue 12, pp. 78–82.

[8] King, Rachael, "Show Time," *tele.com*, Vol. 3, No. 2, pp. 78–82.

Part IV

The Product

9

Branding

Defining a brand

Any discussion of branding should include several definitions. Branding refers to the way a company creates recognition and an appetite for each of its products in the marketplace. Effective branding has two important characteristics: a name, symbol, or slogan that customers use to identify the product or service and a customer perception that the product is in some way different from those of its competitors. The brand is an asset, similar to company plant, cash, and receivables. As such, it has asset value, sometimes referred to as brand equity.

Telecommunications providers can *directly* control the name, packaging, symbols, and slogans associated with their products. However, they have only *indirect* control over the perception of their brand held by customers. A telecommunications service provider is limited only by imagination and budget to create a brand—through advertising, quality control, distribution, and customer service—but, in the end, the brand image exists only in the customer's mind.

While a company's brand does not need to be different in any actual or physical way from its competitors, a successful brand does need to be perceived as unique by customers. Branding enables a company to command a higher price than an equivalent unbranded product. This premium is often called the brand tax or the brand premium. Therefore, companies with strong brands are more profitable than companies without them, if the brand premium was obtained in a cost-effective way.

Not all products enjoy the visibility or the benefits of a brand. A commodity is a product that is exactly the same in physical and other characteristics as those offered by any other provider and that is viewed as identical by buyers. Commodities include crude oil, farm products, salt, and dirt. Commodity products generally sell at low margins. Commodity buyers are extremely sensitive to differences in price, as customers see few differences in the product or the way it is delivered. Therefore, companies do their best to create a brand image for their products.

For the most part, today's local and long-distance telephone services are also commodities or near-commodities. Local service is a commodity because of historical pricing and regulatory policies, which made certain that all telecommunications users were treated equally. Long-distance service is becoming more of a commodity because service quality is similar from all suppliers and because pricing has been simplified. Internet access, unless differentiated by access speed, access points, or content, is a commodity. Telecommunications products resemble commodities because they have not needed to be differentiated. Competition will change this.

Positioning the brand

Part of a brand's identity is its positioning. Customers perceive any brand to have its own qualities, such as its effectiveness, reliability, or price. Positioning occurs when customers observe the attributes of brands and see the brand as it compares to its competitors. A telecommunications provider can try to influence the position of its brands in customers' minds by comparing them to other products. The Sprint "pin drop" campaign and AT&T's "True Voice" advertising are attempts to position the products based on network quality. A product's positioning can rely on its price, the benefit it offers, the market it serves, or its distribution. When a company formulates the product positioning, it needs to recognize that positioning a product in one way eliminates other positioning options for the brand. Thus, if a provider positions a product as the low-cost alternative, the same brand cannot be positioned as the luxury alternative.

Some companies actually succeed at positioning similar products as separate brands by changing their attributes as perceived by potential customers. In the relatively new competitive world of telecommunications, examples do not abound. Historically, though, local service providers offered single-line business or residential local telephone service at widely variant rate structures, as required by regulators. In a competitive world, that brand distinction probably will not survive, but similar pricing variances are easy to find in airline pricing since deregulation. For a business flier versus a vacationer, the same seat on the same flight can vary in price significantly, with no more product differences than the day the ticket was purchased or the timing of a return flight. Customers often look for the branded "super-saver" flights when they make their reservations.

Most of the largest telecommunications providers have not committed to positioning their offerings effectively. In fact, several have done the opposite by establishing campaigns to brand their services very generically. Telecommunications giants do not appear to be willing to position their offerings because positioning necessarily limits the markets they will serve. They would like to provide all services using all technologies to all customers in all locations. This strategy makes more sense in a monopoly market than a competitive one. Few companies succeed without targeting a market segment or several segments. (This topic is discussed in more depth in Chapter 5.)

Today's telecommunications branding strategies

Customers find brands of telecommunications leaders everywhere in the marketplace. The brand can include the company name, the names of company divisions, the names of products, and the names of bundled packages. The telecommunications industry lags behind other industries in using brand management as a sophisticated marketing tool. While few telecommunications companies have mastered brand recognition at the product level, they have worked hard to achieve recognition at the corporate level.

MCI's brand strength and worldwide name recognition represent a large investment by the company. Sensibly, the company has abandoned communicating its original name (Microwave Communications, Inc.), which is wedded to a technology that is unimportant to the company's image. AT&T has done the same with the reference in its own name to telegraph, clearly an insignificant if not extinct segment of its business.

Many companies have registered several variations of their corporate name as a Web domain. A user can reach Bell Atlantic using www.bell-atl.com or www.bellatlantic.com and separate domain names for subsidiaries. The more registered variations, the more likely potential customers can find the site easily. Moreover, companies further protect their names by registering Internet domain names that could parody or be confused with their own sites. Companies created in the future will need to consider not only the immediate impact of names and brands but translating the names to an easy access Web address.

One of the most clever and successful names in the telecommunications industry was given to Bellcore, a contraction of *Bell* Communications *Research*. It contained a play on words, as its owners, the RBOCs, intended the research facility to concentrate only on "core" business interests. Unfortunately, its new, non-Bell ownership and the expansion of its mission have forced a name change. The company's apparent strategy is to use both its old and its new name, Telcordia Technologies, until the new name is well recognized.

Companies guard their brands intensely, through patents, trademarks, and service marks. They are correct to take legal action against those who either steal or damage intentionally the brand equity they have built. Most of the substance of a brand is its image; most of the value of the brand is inside the customer's mind. Anyone with a will and the means to hurt the brand or company image can reach the mind of the customer.

While companies try to gain corporate recognition through publicity, advertising, and sponsorships, their efforts at product branding are not as apparent. When telecommunications existed in a monopoly environment, managing brands, like most marketing functions, did not have to be sophisticated. A provider in a monopoly market can invest in a market-building advertising effort, as most of the new market will be theirs. AT&T's famous "Reach Out and Touch Someone" campaign was focused on enlarging the profitable long-distance market. At the time, there was little or no significant competition for long-distance services.

As AT&T's market share drops below 50%, the company has become more resolved to develop and nurture a powerful brand identity for the corporation and its products as distinct from the long-distance market as a whole. AT&T has been very successful, sometimes more successful than its actual market presence. Customer inertia, in addition to loyalty, accounts for some of AT&T's market leadership.

AT&T remains the strongest company name for telecommunications services. Two-thirds of 800 survey respondents rated AT&T as "excellent" or "very good" as "a telecommunications provider of the future" [1]. This is good news for a company that plans to expand its product line and its geography. For local providers, name recognition is strongest in the existing territories. The administrators of the same survey concluded that the name Bell contributes to name recognition.

Even a decade after giving up the local service market in its divestiture from the regional Bell companies, AT&T was named by 30% of consumers as their local provider [2]. In a more recent survey, only 5% believed that AT&T was their local service provider. At that time, AT&T had very few or no residential local service customers.

Many telecommunications providers do not have a coordinated strategy for branding and fail to gain additional sales or market strength from their brands. Others try to extend their well-known brands to facilitate entry into new markets, a strategy that does not always work. For instance, neither AT&T nor IBM has become a leader in the Internet access market, even though both enjoy valuable brand identities in various other technology markets. Companies can also weaken brands by extending them to new lines that undermine the images held by existing brands.

Brand management and name recognition

Brand management involves developing and communicating the uniqueness of products, their delivery, and the reputation of the parent company to the marketplace over the long term. A strong brand is a vital corporate asset that can have significant impact on the success of a company. AT&T, US West, and other telecommunications providers have recognized the importance of protecting their highly regarded brands and established organizations for brand strategy and marketing communications.

For more proof that telecommunications providers are struggling with brand recognition, note the following study. A survey asked 500 residential customers what they want from their telecommunications providers [3]. When asked to name a local provider, nearly 30% named AT&T. Motorola, a manufacturer of wireless products, was named as the third most popular cellular provider. These two respected companies actually hold name recognition in markets they do not presently serve.

These results are not unilaterally bad news. Certainly it will be easier for AT&T to enter local telecommunications markets because customers already believe that it is there. Nevertheless, its modest success in Internet access is proof that leadership in new markets is not guaranteed.

The company name holds much of the brand image for most telecommunications companies, as their products, so far, are not very distinguishable from each other's. After Bell Atlantic merged with NYNEX, the surviving name was the one that retained the Bell brand, and the geographical reference remained accurate. The merger between Southwestern Bell and Pacific Telesis kept the Bell reference as well, although it has been compressed into the SBC acronym. The MCI-WorldCom merger inspired industry observers to wonder which company name would survive, as MCI boasted more worldwide recognition for the brand but WorldCom invoked a more accurate and futuristic product image. The combined entity currently operates as MCI WorldCom until the combined brand is perceived as having meaning to customers. This strategy will enable the company to develop the brand further and eventually settle on one of the two name elements or a new identity. It is possible, though unlikely, that the interim name will be permanent.

Companies with a strong product strategy sometimes avoid using their most recognizable brands to create a new image. Suppose a company with an image for high-quality networks wanted to offer a low-cost brand. Using a brand extension for the low-cost product could tarnish the value of the high-end brand.

MCI has spent most of its first two decades trying to convince customers that it is a worthy competitor to industry leader AT&T in quality and customer service, with a small discount in its price. Then MCI decided to enter the low-cost market. Its low-cost 1-800-COLLECT and Telecom USA services are designed to convey an off-price, wholesale image that is inconsistent with the industry leader image that MCI wants to sustain in most markets. No advertising for these two products mentions the corporate link with MCI. MCI does not want the customers for these low-cost services to believe that they are associated with an industry leader. Clearly, MCI wants customers to believe that they are enjoying very low prices outside the mainstream of carriers. AT&T has done the same with its off-price long-distance brand, Lucky Dog.

The brand in the business market

While much of the research surrounding brand recognition is conducted among consumers, the business market is equally responsive to a strong brand. The phrase, "Nobody ever got fired for buying IBM," demonstrates the brand power that was once held by the mainframe manufacturer.

Businesses often can be less price-sensitive than consumers can. Frequently, they have access to more funding, and a product that businesses purchase can be justified through increased revenues, reduced cost of ownership of the product, or duration of the product life cycle. Businesses, more than consumers, can look beyond the inherent physical quality of the product. A technology-driven supplier that does not build a well-rounded brand image that appeals to the business buyer will not perform as well in the marketplace as a less-qualified product with a better overall structure. Apple Computer is perhaps the best known example of this situation in the personal computer market.

Achieving brand recognition matters only in the markets to which a company wants to sell. For example, the leading brand of fiber-optic cable is probably not known by most of the world. It does need to be well known by those managers who buy fiber-optic cable. Part of brand management, especially in the business market, is to identify the universe of potential customers and create a program that communicates to them.

The benefits of a brand

Why does a telecommunications company want to create a brand to sell to customers? Brands enjoy longevity. Twenty of the 25 leading brands in 1923 were still in first place in their product categories six decades later. Four were in second place. Well-known brands last and provide profit and corporate value for years. It has been estimated that half of the market value of the Coca-Cola Company is the ownership of its brand.

Brands engender loyalty. Customers find it easier to buy a well-known brand, and corporate purchasers find it much easier to justify procurement of a known brand, even overcoming price differences. Once the customer owns a branded product, it is harder for an unknown competitor to unseat the branded product. With all the discounts available on long-distance services, about 60 percent of households still paid nondiscounted rates more than 10 years after divestiture and nearly 20 years after competition was introduced [4]. The most plausible explanation is brand loyalty. This loyalty is somewhat misguided, since the very same carrier would be willing to offer the discount to any customer knowledgeable enough to ask for it.

Brands ease the way for new related products. Companies with successful brands have an easier time selling new technologies and new products to customers. This technique has a dual edge, though. The company with a reputable brand that delivers a failure of a new product runs the risk of devaluing the strength of the existing brand.

Brands mitigate the consequences when a company's market entry is late. It is often argued that the personal computer (PC) market did not really develop until IBM released its PC. The phrase *PC-compatible* still demonstrates that the IBM brand brought credibility and a technical standard to the market. While IBM was certainly a late entrant to the PC market, it managed to claim and keep a very large market share.

Not all brand strategies succeed

Most brand management specialists agree that a brand strategy is a corporate, comprehensive program, supported by external communications, product development, and all other marketing efforts of the company. Often, the telecommunications industry has lagged behind more customer-driven industries in branding specific products. While it has named various services and customer packages descriptively, the lack of competition and the commodity nature of the more competitive services has failed to provide the requirement for differentiation through branding. In other words, telecommunications product names have traditionally been boring. Services were called "the Yellow Pages," "extended area service," or "wide area telephone service (WATS)." Many commonly used telecommunications product names were derived from engineering or regulatory references. Routinely, companies include regulatory and technical terminology in their advertising and their bill inserts. Generally, these terms are not very meaningful to the average customer. The fact that customers are exposed to the back-office infrastructure is an indication that these companies are not doing everything they can to control their brand images. Marketing-driven companies would coin more dazzling terms to exploit the benefits of new services offered through regulatory or technical improvements.

Branding entails more than creating a name for a product or service and advertising it. Ideally, brands should be developed as a result of market research that concludes that customers would be receptive to new products that share the following characteristics:

- They do not presently exist.

- They can be produced economically and sold at a profit.

- They can be produced without excessive disruption to existing operations.

- They will not unduly cannibalize the existing product line.

In technology industries, companies often need to create new products and brands, even when they will erode existing markets, to hedge against competitive inroads.

Several telecommunications providers have created brands before the anticipated product or service was technologically available. MCI has tried to develop brands for complete telecommunications packages, such as its networkMCI One umbrella for services. Its success is not yet proven because, like other providers, MCI is not yet capable of providing truly comprehensive solutions to all customers. AT&T created a brand called PCS while the marketplace was waiting for generically labeled personal communications services (PCS). The PCS service offered by AT&T, while digital, did not strictly adhere to the industry's expectations for the technology. AT&T's introduction of the new service was an attempt to be first in the marketplace and create a customer base for its eventual digital offering. SBC's efforts at branding were criticized after it unveiled a major positioning strategy and then was unable to elaborate on the strategy and how it fit SBC's overall branding efforts [5].

Like most industries, telecommunications giants have used acquisitions to provide growth. In some cases, the strength of the acquired brand(s) has helped to justify the acquisition. AT&T's acquisitions of the McCaw and NCR brands did not provide lasting brand equity (although the McCaw acquisition provided needed wireless infrastructure and a customer base).

It is probably fair to state that there are no spectacular product brands in the competitive telecommunications service world yet. Perhaps the closest anyone has come is the "Friends and Family" program offered by MCI in the 1980s. Generally, telecommunications branding for local service offerings has suffered from the inexperience of the marketers. ISDN has been generally a failed brand for each of the local exchange companies that have tried to offer the service to consumers. Among its marketing problems was timing (high-speed access was unimportant before the Internet was ubiquitously used by consumers), the difficulties and costs associated with its implementation, and its pricing structure. Another of the many problems that faced ISDN was its name, which was hard to pronounce and told the customer nothing about the benefits of the service or the problems it solved. A legacy of the facilities-driven monopoly era, the brand ISDN merely described the technology underlying the service in a most generic way.

Turning commodity products into brands

A brand needs to be perceived by customers as unique in comparison to its competitors. Commodities generally cannot do this because they are fundamentally equal to their competitors. The most common way to differentiate and sell commodity products and services is through lower pricing rather than creating a unique image or brand. Most companies dislike this low-price strategy, as commodity margins are not high in the first place. Reducing prices invariably leads to price wars (witness the airline industry after deregulation and to a large degree the long-distance market). Companies in price wars do not make short-term profits, and they create unrealistic price expectations in their customers, which hurts their long-term profitability.

Ironically, some of the best regarded telecommunications products (e.g., central office-based voice mail services, long-distance, and satellite television) are priced well above their actual costs and have significant competition or substitutions in the marketplace. Commodity pricing will not be necessary in a competitive telecommunications marketplace, and telecommunications providers will have substantial opportunities to build brand image to command higher profits. Whether low price can be an effective means of competition is dependent on the provider's cost profile. The relationship of costs and prices will be covered in Chapter 11.

Instead of price wars, companies prefer to differentiate their products, as it is the primary means for branding products. Once the product is viewed as different by the customer, the brand serves to sustain the differentiation and associate it with the provider. Branding and differentiation can be based on the inherent characteristics of the product, as Sprint attempted to accomplish with its "pin drop" advertising campaign. Its goal was to convince customers that the quality of the long-distance connection was technically superior to that of other carriers, targeting AT&T and MCI. While the "pin drop" campaign did achieve some name recognition for Sprint outside its franchised local service areas, the "pin drop" profile was later reduced substantially when the company opted to differentiate by price (the dime-a-minute campaign).

It is not necessary for branding and differentiation to be based on the physical characteristics of the product. In fact, for commodity products, it is often well worth the effort to differentiate based on factors other than quality or price. In the telecommunications industry, quality is generally so good that it would be price prohibitive to achieve leadership in technical quality. Instead, telecommunications companies should be expected to differentiate in areas such as customer service, system coverage, and product delivery.

A recent J. D. Power study [6] determined that consumers value factors other than call quality for long-distance services. See Figure 9.1.

In fact, the value of the brand itself, that is, the image of the provider, was more important to customers than the quality of the call. Note, however, that this study only considers long-distance service, and call quality can be perceived by customers to be a function of both the local and long-distance provider. For most customers, local and long-distance services are offered by separate companies. Note also that customer opinion about call quality is not the same as the actual call quality and that the customer

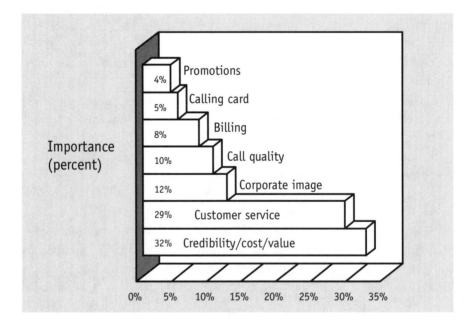

Figure 9.1 Factors affecting long-distance selection.

requirements as reported on surveys do not necessarily drive their purchase decisions. For high-volume users (over $50 per month), the highest scoring company was Sprint. It would be impossible to judge the effect on these scores from the "pin drop" campaign. Did actual quality contribute to an overall customer assessment of "credibility/cost/value"? Did other factors help Sprint in the high-volume segment?

A branding study [7] conducted on behalf of publisher Ziff-Davis found that five factors affected the high-technology purchases of consumers and businesses. All of the vendors reviewed were manufacturers of computers or system components. The factor analysis yielded five essential factors that drive the purchase decision: price/value, customer focus, user friendliness, leadership, and Internet support. For the most part, these attributes are achieved outside of the manufacturing facility—in research, customer support, and market analysis. Telecommunications providers can expect similar demands in a fully unregulated marketplace.

Innovation and competition

MCI has historically been an innovator in the delivery of services to consumers. MCI has offered a discount to long-distance customers who wish to pay their monthly bills on a credit card. By using the service, called MCI One Net Savings, customers can review their bills on the World Wide Web and no longer need to write a check each month. MCI reduces its administrative costs (other than the credit card fees) by eliminating the need to mail a monthly statement and manage the receipt and deposit of checks. Simplifying bill payment should also serve to reduce uncollectibles. While other companies can respond competitively and copy the offering without too much investment, MCI obtains the advantages associated with being first. When its competitors eventually offer the same service, presumably at the same price, there is no reason for the current MCI customer base to change carriers. The feature was suspended when MCI's Internet backbone was sold to Cable and Wireless. Still, this offering is one of several examples of MCI using its superior information infrastructure to gain and retain customers. Others are discussed in Chapter 18.

Product differentiation is ideally based on some attribute that competitors cannot match in the short term. While this is not easy, superior marketing in telecommunications has produced some success stories. For example, Tel-Save, a discount long-distance carrier, entered a contractual relationship to advertise and sell its service on the America Online (AOL) home page. AOL subscribers can sign up through a series of screens, and the company bills the calls on their credit cards. This provides a competitive advantage to Tel-Save that cannot be matched by competitors. AOL boasts the largest subscriber base of any online provider. Most of its subscribers are consumers rather than businesses, offering built-in customer segmentation to price-sensitive buyers. The service's brand name is clearly targeted to this market segment.

This advertising and sales channel is superior to telemarketing or print advertising because it offers an access to customers that is unavailable to its competitors. If the Tel-Save brand can add value—such as online billing or access to special discounts—to its AOL customers, this can differentiate the product for the long term as well. Furthermore, this represents an enormous strategic threat to today's local exchange providers. Many of them expect to provide local service to consumers outside their currently franchised territories. According to a recent study, the AOL brand is stronger nationally than any of the RBOCs or GTE, except in the home regions of these local carriers. On the other hand, AOL's reputation for uneven reliability might affect its success in the local service arena.

Pricing can provide a differentiating factor, even when the competitive advantage is not simply low price. Historically, long-distance pricing depended on many variables, including jurisdiction, distance, time-of-day, and types of customer. Customers were accustomed to paying a multiplicity of per-minute charges, and it took several years for providers to realize the marketing benefits of a simple pricing structure.

The price structures of the last decade have become simpler—but slowly. When the first flat pricing was introduced, Sprint customers who signed up for the heavily advertised "dime-a-minute" price did not necessarily realize that the low price was time-sensitive. Some carrier programs require a certain level of monthly calling before the user is eligible for the program. Other programs require a monthly cover charge before any calls are made.

AT&T is attempting to capitalize on this and pricing strategies of other carriers by offering a single price to consumers for virtually any domestic call at virtually any time. The AT&T One-Rate campaign is intended to eliminate the confusion that customers have in comparing one service to another. AT&T never advertised that it offers the lowest price and apparently does not intend to compete solely based on price. The company is betting that consumers, tired of wondering which complicated pricing scheme works best for them, will be drawn to the ease of AT&T's offering.

Until the end of 1998, no carrier challenged the concept of timed long-distance calling. Differentiating through flat monthly long-distance is still a rarity. When Internet-based telephony becomes a significant competitor, and subsidies are removed from the long-distance cost structure, the cost-per-minute could drop substantially. If that happens, more providers will offer an unlimited calling package to the highest-volume customers and continue to earn a profit on the service.

While product differentiation is a way to develop a brand perception, real product differences are not required of a successful brand strategy. A strong brand by itself and not simply a product's inherent differences can help a company to gain a higher price for the product, everything else being equal. The company's challenge is to develop the brand awareness in the customer based on true added value and reap its benefits over the long term when other providers' products have caught up. The premium of the brand, which can reach 15% of its price, can last much longer than the actual product differences.

In some ways, this demonstration of the brand premium is a something-for-nothing concept similar to the accounting concept of goodwill. It covers the higher price or purchase decision directed toward a product when there is simply no other good reason for the buyer to make that decision or spend the additional money.

SELF-ASSESSMENT—BRANDING

Telecommunications marketers should ask the following questions about their own brands and brand strategy.

- What is your company name? Do prospects other than your current customers recognize the name? What is the message conveyed by the name? Does it limit the customer image to a technology or geography?

- Does the company conduct the same business under different names? Is there an important strategic or brand positioning reason to do this?

- What is the company image to your targeted market segment? Does the image held by customers and prospects match your strategic and marketing goals?

- Do the names of products support the company name and image? Are product packages and names planned from the view of the corporate branding strategy? Does the advertising program support the corporate brand?

- What do your prospects, not your customers, think of your brand or brands? How often do you study this?

- Do your employees understand your brand strategy?

- How strong are the brands of your competitors? Is there a way to exploit the flaws in the images of competitors' brands?

- What is the effect of the marketing channels (your company's representatives) on the brand perception? Does the channel support the brand perception? Does it make sense to create new brands for different marketing channels?

References

[1] Meade, Peter, "Is Bundling Really Better?," *America's Network*, Vol. 100, No. 18, p. 26.

[2] Morri, Aldo, "Carriers Lament Customer Confusion," *Telephony*, Vol. 233, No. 3, p. 50.

[3] Egolf, Karen, "Voice of the People," *Telephony*, Vol. 230, No. 12, p. 38.

[4] Mills, Mike, "Missing Out on Lower Long-Distance Bills: Discount Plans, Higher Rates Part of Telecommunications Debate," *Washington Post,* Final Edition, Section A, p. 1, July 2, 1995.

[5] Salak, John, "The Big Brand Era," *tele.com*, Vol. 2, No. 8, pp. 76–79.

[6] J. D. Power and Associates, "1998 U.S. Residential Long-Distance Customer Satisfaction Study (SM)."

[7] Ziff-Davis, "ZD High-Tech Branding Study," conducted by IntelliQuest on behalf of Ziff-Davis, August 1997.

10

Service Development

The product/service development process

Service development is a crucial marketing concern in a competitive environment. New services reach new markets, generate interest and loyalty among existing customers, and increase market share. The process is ongoing, expensive, and less scientific than most providers would like it to be. As a starting point, Figure 10.1 depicts the activities required in the development of new telecommunications services.

In a deregulated market, virtually everything about service development is different from monopoly traditions. Under monopoly, service development, like all other functions, is driven primarily by a requirement to provide high-quality, inexpensive basic telecommunications service to the largest universe possible. Monopoly structures around the world did indeed provide affordable, reliable service for the first century of the technology. Nevertheless, nearly every business incentive and objective will be transformed in a competitive environment.

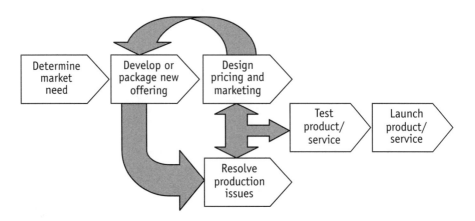

Figure 10.1   The service development process.

In the process chart shown in Figure 10.1, the development process is driven by market need, not by an array of regulatory objectives. New services can include the following:

- Improvements or repackaging of existing services (such as unbundling or bundling services);

- Discontinuous innovations such as digital subscriber line technologies;

- Dynamically continuous improvements such as digital cellular service instead of analog service;

- Continuous innovations such as upgrades to reliability.

The following sections demonstrate why the development and introduction of services in a highly competitive market will be very different from their counterparts under a monopoly system.

Nature and timing of service introduction

The first change is that no telecommunications provider, including those with internal product development capabilities, will be able to control completely which services they introduce and when they introduce them.

The successful telecommunications provider will need to address the pressures exerted by both customers and competitors.

The concept of time to market emerged because the rise of technology in most industries shortened the development cycle. Customers demand new products to maintain the continuous stream of recent history. Computers and telecommunications equipment double in capacity every 18 months, a rate with no signs of abating. Customer bandwidth demands double at the same rate.

Companies that are first to market enjoy no competition for a limited time, which permits them to charge premium prices and gain profitability. Where legal and feasible, the companies that copy and offer the new services soon thereafter (through reverse engineering or substitute technology) gain much of the market advantage but incur little of the cost. Each telecommunications provider will need to make a strategic decision as to whether it wants to make sufficient investment to maintain technological leadership. Not every successful provider will be a technology leader or first to market. Conversely, not all technology leaders will be successful. The race to market can cause providers to launch services before the technology is reliable or rush new services to market when the potential customer base is uncertain.

In the U.S. telecommunications monopoly, the local service provider, with the cooperation of the regulatory authorities, controlled the nature and timing of product introductions and innovation. The technology for cellular mobile service was available two decades before the first service introduction in the United States. A variety of delays prevented the launch of the service, most of which related to regulatory issues. Customer demand was apparently not as influential as the regulatory and administrative issues affecting the cellular rollout.

Prior to AT&T's divestiture of most of its operating companies in 1984, Bell companies purchased all of their major network equipment from their parent company. Independent local companies utilized other suppliers, but the Bell system was historically the innovator for the implementation of new system-wide services. For the most part, services that involved multiple providers, such as the automated credit card of the early 1980s, were introduced by AT&T, and the non-Bell providers cooperated in marketing, billing, and operations.

The most important aspect of this monopoly pursuit of innovation is that a single provider controlled virtually all aspects of the service introduction, including timing. Service launches were timed to meet the best interests of the service providers and the regulatory view of the customers. Telecommunications providers never needed to introduce new services before there was enough capacity, before the administrative infrastructure was in place, or in a nationwide rollout. Most new services were tested in limited market situations before they were launched to a widespread customer base. In a competitive market, this will not be possible.

The goal of affordable universal service also affected the introduction of new technologies. In general, electronic switches were incorporated into the network infrastructure at the rate that the older mechanical switches were no longer useful. Mechanical switches wear out and have a defined capacity. Electronic switches are physically usable for a good deal longer than their market life span. An electronic switch can operate without failure for decades, even if an upgraded switch is available in the market after just a few years. It is common that electronics, including computers and telecommunications equipment, become obsolete well before they stop working.

Depreciation is the accounting practice of deducting part of the price of a piece of equipment each year when the equipment lasts more than one year. The purpose is to match the initial cost of the equipment to its useful life. In the interest of holding subscriber rates low, depreciation lives for network equipment were historically set to be as long as possible. The expense for a piece of equipment, recovered through subscriber rates, is half as high per year when the useful life is twice as long. Depreciation rates were most often based on how long the equipment would work, rather than the market's need for new services.

Telecommunications providers in the 1980s, in light of new switching equipment with enhanced capabilities, faced an unappealing choice. They could delay the introduction of new services to their customers or abandon usable network infrastructure before its costs were recovered through depreciation.

The practice of controlling innovation will decline in a competitive market. Customers will demand new services, more bandwidth, and cheaper technologies. They will buy services from whatever provider offers the services they desire at a price that they consider worthwhile.

Innovation and profitability

The monopoly environment, coupled with the vertically integrated telecommunications provider, gave providers nearly free rein to conduct research and hope for breakthroughs. If research conducted under the monopoly umbrella led to new services that could provide substantial profits, the services were priced much higher than their operational costs and perhaps above their development costs. Patents were not as important as subsidies and monopoly for protecting the income stream. AT&T's transition from primary telecommunications provider to market competitor forced management to make tough decisions concerning the price and scope of its world-renowned research capability.

Simply stated, monopoly pricing recovers all costs from all services, in some proportion determined by regulators. No single innovation or research path required justification based on its market success. Competition does not permit this luxury. While some research is still heuristic, competitive telecommunications providers are applying more limits on the type and amount of research they fund.

While the world's carriers have not reduced their research and development budgets (about 2–3% of total revenues), they are moving away from heuristic research and toward research with more practical applications [1]. AT&T focuses internal researchers on product development and uses university affiliations for longer term projects. Both AT&T and BT allocate about 10–15% of their total research budget to pure research.

Telecom Finland was one of the first telecommunications providers to experience significant competition. The provider has decentralized its research and development operations to the business units, where teams work alongside customers [2].

Because the pricing of new services is market-driven in a competitive market, companies cannot use service introductions as a long-term source of profits. For one thing, substitute services offered by competitors place a cap on the price that a provider can charge for most new services. In addition, because product life cycles are short, the high level of profit will not last as long as a provider would want it to persist.

From averaged to customized offerings

Monopolies are optimized when they achieve economies of scale. These economies are always present in monopoly markets such as telecommunications or utilities; they provide the justification for "natural monopoly" regulation. The telecommunications industry was certainly characterized by economies of scale in its earlier days.

One artifact of these economies is that services offered to customers were almost identical. Pricing was a different story, based on meeting regulatory objectives. The service itself was averaged. In the days of mechanical switches, one access line was nearly indistinguishable in function from any other. The list of service offerings was short. This arrangement was by design. Offering millions of a single service is less expensive and carries less risk than offering thousands of thousands of services. Customers, if asked at the time, were probably quite satisfied that their telephone service was limited, especially if the limitations kept its cost low.

Electronic switching introduced a variety of new services and created an appetite for customization among customers. The enhanced services available through electronic switching and a more intelligent network also provided revenue opportunities and more subsidies from customers who could afford them, pleasing providers and regulators. Competition will raise the stakes on customization of services.

Competition will require providers to improve their understanding of the needs of smaller and smaller customer segments and then tailor their services to meet those needs at a competitive cost. Customization will cost more for service development, including more cost at each stage of the development process. Market needs assessment will be more expensive because the customer base will become more complicated. The variety of new offerings will increase the workload and cost in the development, packaging, pricing, and marketing of the customized services. Production issues will become more complex.

Testing will require more subjects in each customer segment and more sophistication to ensure that subjects are inured to competitive assaults. Launching customized products requires more targeted marketing and potentially higher overall cost. While de-averaging the service line appears to be unnecessarily complicated, providers are aware that competitors that offer a service with a better fit to the individual customer's needs will win the customer in the end. Service averaging is sustainable in a market only when competitive, customized services are not available.

Vertical integration

Many successful telecommunications service providers will choose not to be vertically integrated for service development in a competitive market. Some will decide that development is not as important as building infrastructure in new markets; others will appreciate the flexibility of using outside developers; still others will focus on their core competencies.

In the regulated monopoly market, the largest carriers were vertically integrated. They controlled the entire distribution chain, and they acted as their own suppliers. AT&T owned its manufacturing arm, Western Electric, and its research and development company, Bell Laboratories. Other large providers such as GTE and Bell Canada were similarly structured. Smaller U.S. providers that did not develop their own services depended on market leaders such as AT&T to create the technologies and offer them nationwide.

The competitive environment will require telecommunications providers to revise their service development processes. The vertically integrated structures that were once in place are crumbling. The Bell companies sold their interest in Bellcore to have more control over their own companies' research and development efforts. AT&T divested its manufacturing arm, Lucent Technologies, partly because of channel conflict.

If vertical integration enables providers to control innovation, how will providers introduce new services without it? The lack of an internal development capability will not eliminate the need for a stream of new products. Telecommunications providers that choose to offer leading-edge services will need to find sources of innovation and will need to invest in companies that will create new services, whether or not in a proprietary arrangement. Competition will push the level of investment upward for most providers, and the level of aggregate investment will be decidedly higher than it was in a monopoly environment. The result of the added investment will be more products at a faster rate.

Providers will need to recover their investment through increased sales. This will be more difficult than in a controlled, vertically integrated environment. Shorter product life cycles will require faster cost recovery, while competing products will maintain a downward pressure on prices.

Alliances with other providers will enable telecommunications service providers to leverage their development investment without bringing the development capabilities in-house. In exchange for the lowered cost of investment, providers that share development lose some of the differentiating features resulting from their research. When partners are not local competitors, the tradeoff is worthwhile. After making an investment in MCI, BT asked MCI supplier Nortel to recreate the Friends and Family program for BT customers.

Outsourcing development provides many of the benefits of in-house development at a greatly reduced cost. One method is to outsource proprietary service development to a firm whose core competency is development. Carriers need to control the temptation to develop very detailed specifications because the cost can be prohibitive and the rapidity of product obsolescence does not tolerate the long lead times.

Another method is to license the technologies developed by outside firms. This will cost less, but the nonproprietary technologies will be available to competitors, so they cannot serve as a competitive differentiator. One advantage of outsourcing or licensing is that day-to-day management of the entire development process can be delegated, and management can focus on the development strategy rather than its operation.

One result of the uncoupling of service provider and service developer is the emergence of standards. Standards increase a service provider's flexibility to mix the offerings of several suppliers. Standards also facilitate the rapid development that is necessary in a changing marketplace. One service developer claims that standards can reduce development time to two or three months, compared to the nine months a custom solution would take [3]. In any case, the route to developing standards can be serpentine and arduous. While standards facilitate development, waiting for standards can be just as frustrating as combining incompatible applications.

Telecommunications Management Network (TMN) is a model developed by the International Telecommunications Union to manage the telecommunications services environment. Its individual components, or layers, are not of direct relevance to marketing, but its existence as an industry standard is very important. TMN sets a standard for service development using an open architecture, and its goal is to provide for solutions that offer interoperability and rapid development. Benefits include improved quality, lower costs, multivendor support, and improved network management.

Convergence

Telecommunications and information systems are moving together in a process known as *convergence*. Some industry observers maintain that deregulation in the telecommunications industry was a significant driver of the convergence; others hold that technological advances naturally drew the two disciplines together. Whatever the cause, telecommunications no longer exists apart from other information technologies.

In an isolated monopoly environment, advances in mainframe computing had little direct impact on telecommunications customers. Monopoly regulation prevented most other industries from offering substitute telecommunications services except in very specialized cases. Eventually, though, the two industries began to merge. Switches began to resemble mainframe computers. Other network components took the form of minicomputers, but these technologies were not visible to customers. Personal computing changed that, especially with the rise of Internet computing. PCs provide a highly visible substitute for traditional telecommunications networking.

Convergence changes traditional service development in several ways. Computing is both a substitute and a complement to telecommunications service. By providing an alternative to traditional telecommunications networks, it enhances the level of competition. By offering progressively inexpensive processing capability, computing created a need for increased telecommunications capability. Convergence will contribute to market growth and create the need for new network features. On the other hand, convergence will compete for the customers of the features that once resided exclusively in the network. Customers who once used central office-based answering systems or purchased answering devices from telecommunications manufacturers are now installing computer-based messaging programs that are less expensive and sometimes more capable than their telecommunications network alternatives.

Another outcome of convergence is that telecommunications customers will expect the same intensity of competition and the same rate of innovation in telecommunications as they experience in computing. Telecommunications customers will expect prices for similar services or technologies to fall over time as they have in the information systems market. For the most part, customers have not been disappointed in the telecommunications markets as they are unregulated so far. Wireless prices have dropped each year and so have prices for long-distance and international calling. While local service pricing will be more problematic to reduce, because it is offered below cost already in many locations, a combination of technology, competition, and customer expectations will sustain downward pressure on local service pricing as well.

Network architecture

As in the computer industry, each device or node in the telecommunications network has become smaller and more powerful. Computers moved from mainframes to minicomputers to servers and workstations. Regulated telecommunications networks had a pyramid-like structure in which the largest regional switches handled long-distance traffic and progressively smaller switches made up the local network. For a monopoly serving all customers, this structure offers economies of scale but less flexibility than a more distributed network. Network engineers, in light of technology and market transformations, are revisiting the architecture of the network. The likely outcome is that the local switches will become more powerful and that the overall network architecture will appear geodesic rather than pyramidal.

Providers will need more flexibility in their networks. They will need to turn up services quickly or exit unprofitable markets. They will need to meet competitive offerings in a very short time frame. Smaller, cheaper nodes that are more powerful will help providers offer customized services from central offices. If one resident switch does not offer the desired feature, perhaps another one can. In the distant future, perhaps each customer can configure the equivalent of the local switch and select whatever features are desired.

The first indication that customized features will be available soon is the emergence of programmable switches. Programmable switches process calls and perform administrative functions such as playing prompts and responding to keys pressed. They offer advanced features and flexibility to carriers that do not want to commit to investing in the largest, most feature-rich switches.

The impact that programmable switches will have on product development is still uncertain. Early indications are that this technology will enable smaller providers to enter markets that would otherwise be unavailable to them. Large and small providers can use programmable switch technology to enter low-volume markets such as a rural area with a minimal investment in switching (although the local loop remains a large cost). The software capabilities will undoubtedly reduce the effort involved in adding new services as they come to market.

The rise of software

Another significant change, perhaps more related to technology than to regulation, is the dependency on software rather than the network hardware. The importance of software will be bolstered by the evolution from analog systems to digital systems. In wireless, providers of analog cellular systems could differentiate their products on price, coverage, and other factors. Digital cellular and PCS systems enable providers to offer software enhancements such as messaging services, custom calling services, and other features. IP telephony has the potential to offer the same flexibility. When telecommunications services are primarily software-driven, providers will have more flexibility and capability.

Software development is more manageable than its hardware equivalent in terms of controlling time to market. Hardware involves design, tooling, and production, and the process is sequential. Many software developers now use a parallel development approach, in which teams of software developers work on different components of the same new system, multiplying the resources available (although there are practical limits to this technique). Some software developers maintain three shifts of workers, sometimes around the world. When the workers on one shift finish for the day, other software developers several thousand miles away pick up where the first shift left off.

Software enables companies to introduce new services quickly and eliminate unsuccessful services just as quickly. Most software-based services provide sophisticated record-keeping functions, enabling management to generate reports and analysis of the success of the service.

Some telecommunications providers and their developers are utilizing a software development technique called rapid application development (RAD). This methodology can dramatically reduce the amount of time it takes to develop database applications, if some compromises are acceptable. The concept of RAD is to develop 80% of the application in 20% of the time. The tradeoff is appealing to telecommunications providers that need to launch competitive services in a short time frame. RAD is especially well-suited to changing environments.

RAD includes users throughout the development process, ensuring that users can make their revisions before wasting a great deal of development effort. The speed of the process helps to prevent new services from becoming obsolete before they are introduced. The methodology also uses teams and a set of specialized development tools.

The disadvantage of the RAD technique is that economy or quality, and maybe both, can be sacrificed in the interest of a speedy development process. In a high-reliability environment such as telecommunications services, sometimes these characteristics cannot be risked. Even for these markets, RAD can be used to develop prototype systems to test in certain markets while the higher quality, longer-lead-time systems are under development.

In Europe, BT and more than 100 corporate members created a RAD users' society. One global operator trained more than 100 RAD facilitators and is using RAD in more than 10% of its projects [4].

Fast-track service development is not an end in itself. Service development should be conducted in a strategic context that includes external forces and requirements and internal capabilities. Fast development techniques can support the product leadership of a telecommunications provider, as long as it does not concurrently ruin its reliability and reputation for quality.

Operations support systems development

Telecommunications monopolies always developed and operated operations support systems (OSS) for their own use. When telecommunications providers become wholesalers, these systems are part of the product portfolio. Two separate pressures have appeared on the horizon: the need to support an increased array of applications that change quickly and the visibility of the OSS to customers. Incomplete or unpolished internal applications are merely undesirable. When customers are using these applications, and competitive substitutes are available, it is completely unacceptable. The OSS need to be sophisticated and reliable and serve as a competitive differentiator.

OSS include real-time operational data, automated provisioning systems, customer care management, trouble reporting, fraud management, exception event management, and reporting and analysis. Thousands of OSS products are on the market. Some were developed internally by telecommunications providers, while others are created by software developers that specialize in this new, growing market. Smaller carriers that cannot afford the multimillion dollar development cost have turned to service bureaus to provide their operations support. The relationship between OSS and service development is highly interactive and interdependent. New services require management tools, so they create new OSS needs. OSS drives service development when it enables new services to hatch. Some carriers will decide that OSS capabilities will be a potential point of differentiation, in a market where the network is nearly a commodity. Those carriers will develop OSS as a core competency and carve a leadership niche in OSS-based services.

Telecommunications operational systems are complex and data-laden, and they require considerable reliability. Object-oriented technologies can provide fast and efficient solutions by using database technology and reusable code. Object-oriented technology has been used by GTE, BellSouth, Bell Atlantic, and US West to develop service creation systems to simplify the provisioning process [5]. Development costs can be reduced by up to 40%, and object-oriented applications require less effort to modify when operational requirements change.

Product life cycle

A well-known model of the product life cycle also applies to services. The model contains four stages, the introduction, growth, maturity, and decline stages. Each stage has its own characteristics and marketing strategies. There is no set time limit for each stage; some products and services can remain in a stage for decades, while others pass through them in weeks or months. In the introduction stage, a service is characterized by few customers, low sales, low or negative profitability, and few or no competitors. Strategies in this stage are selective distribution to contain costs and high initial prices to recover investment. Today, satellite-based telecommunications is in the introduction stage.

The growth stage is characterized by a rapid increase in sales, heightening intensity of competition, and high profits in the early stages of competition. Strategies for this stage include adding new distribution channels, targeting more customer segments, and increasing the variations of the product. Internet access is a telecommunications service in the growth stage.

Attributes of the maturity stage include intense competition and a flattening of sales. Profitability becomes lower in this stage. Strategies are intensive distribution, line extensions to refresh the product line, and high levels of advertising and promotions. Long-distance services in the United States are in the maturity stage.

The decline stage is characterized by declining sales, few loyal customers, and profits at the lowest levels of any stage. The strategies for this stage include reducing new investment, harvesting remaining revenues, consolidating lines, discontinuing products, and exiting markets. In the telecommunications market, there are few examples of declining services. Services will decline when they become technologically obsolete or when new technologies make the older technologies less cost-effective.

The life cycle model is essential because it identifies the vulnerabilities of the existing service line, and it provides a blueprint for a marketing strategy as services pass from one stage to the next. Telecommunications providers will benefit from evaluating each service positioning in its life cycle during the marketing planning process.

SELF-ASSESSMENT—SERVICE DEVELOPMENT

Answers to the following questions will help telecommunications marketers to improve the service development positioning of their company.

- Does your company have a service development strategy? Does the strategy include technology leadership? Does your company's level and type of investment support the strategy?

- Is your company vertically integrated? Is vertical integration necessary in the context of the overall company strategy?

- Are services customized or averaged across the customer base? Is the present array of services vulnerable to competitors offering more customized services? What actions have been taken to head off these threats?

- Has the service development strategy changed substantially as the market structure changes? Is there a mechanism to revisit the service development strategy within the strategic planning process?

- Will your company's scope support the desired level of service development? Does development reside inside the company or is it outsourced? Does your investment in proprietary technologies match the requirements of the service development strategy?

- Has your company utilized faster service development techniques? Are the techniques appropriate within your company's strategic direction?

- Does your company conduct life cycle analyses for all services in the portfolio? Is there an acceptable mix of services in the various stages of the life cycle? What actions are routinely taken to ensure that the mix remains satisfactory?

References

[1] Evagora, Andreas, "The R&D Factor," *tele.com*, Vol. 1, No. 9, pp. 93–95.

[2] Evagora, Andreas, "The Nordic Track," *tele.com,* Vol. 3, No. 1, pp. 68–72, 75.

[3] Robinson, Brian, "The Incredible Shrinking Development Cycle," *tele.com*, Vol. 2, No. 6, pp. 109–111.

[4] Forge, Simon, "Reshaping Tomorrow's Telco," *Telecommunications*, International Edition, Vol. 30, No. 8, p. 30.

[5] Negrete, Ricardo, "Time to Market," *Telephony*, Vol. 230, No. 19, p. 44.

11

Pricing

Pricing strategy

Deregulation has the potential to offer the cleanest slate in history for set-ting prices for telecommunications services, but early indications predict that it probably will not. Too many legacies, social policies, and customer expectations are likely to prevent a genuine market-based pricing structure.

In the monopoly environment of the United States, telecommunica-tions providers created a network that would meet the total require-ments of their awarded franchise at an agreed-upon level of reliability and quality. They then added up their costs of providing services. They sub-mitted these costs to regulatory agencies, which attached a rate of return.

This large aggregated number was then divided up into rates for vari-ous classes of subscribers such as consumers and businesses. Anyone who could afford the luxury services of the time, such as long-distance or cen-tral office-based enhanced services, was charged more than a normal market cost. When local subscribers in an area could not be asked to pay the entire cost of the service, assistance in the form of subsidies came from all around the country. Stated simplistically, all of the revenues paid for all of the costs, but the relationship between the costs and prices of specific services was extremely muddled. No competitive markets enjoy the flexibility of pricing in this manner.

While the telecommunications industry is struggling toward deregulation, public policy considerations have not disappeared. Rural customers still demand that their low rates be maintained, and government mechanisms will ensure that they are. Perhaps technology advances and lowered costs will eliminate the need for rural subsidies someday, but competition among providers is fighting to lower prices at the same time. A new outstretched hand has emerged as well; the desire to provide Internet access to libraries and schools is creating an extra subsidy requirement on the traditional local service bill. One complicating factor is U.S. government policy, which continues to send strong signals that Internet access will not be taxed. When technology enables facilities-based Internet providers to add local and long-distance telephony to their product lines, the question of subsidies through public policy will be challenged. In the meantime, some customers pay surcharges and high prices to telecommunications providers and the government, and other customers pay different surcharges and lower prices than they should.

Pricing any products or services requires a reconciliation of competing conditions. Pricing telecommunications services involves at least the considerations described in the following sections. Many factors contribute to price levels and price structures. The weight given to any of the following issues depends on the industry, the company's situation, and the overall economy, at the very least. Pricing is a complex and indistinct science. Some of the following factors tend to raise prices; some tend to apply downward pressure. Ultimately, all require attention.

History and customer expectations

Under total regulation, prices are set by an agency. Prices are nonnegotiable by the customer and are often based neither on the cost of providing service nor on the availability of alternatives. Generally, alternatives are not permitted. Providers in a deregulated marketplace need to develop prices that will persuade customers to buy. In the early phases of deregulation, telecommunications providers have few experiences to draw upon to help them design the structure and level of prices for services.

Monopoly telephone pricing in most countries is based upon social policy instead of costs. This results in a set of customer expectations about how high prices for certain services should be and whether prices are flat-rate or usage-based. These structural differences do not reflect the actual costs of the service, but customers believe they do after decades of paying for them. Therefore, it is not completely at the providers' discretion to set price levels and price structures.

Customers have preconceived notions about the price and value of telecommunications services. Customers have historically complained about the level of local charges, more than they have about long distance, although local service is frequently offered at a price lower than actual cost. When long-distance service is priced well over cost, and local service is generally priced well under cost, customers expect to pay very low prices for local services and apparently do not mind that long-distance could be less expensive but is not. Perhaps these customer reactions are based partly on their ability to control the expenditures on long-distance calling because the charges are usage-based.

Customers are accustomed to some flat rates and some usage-based rates. Similarly, customers in the United States expect to pay a fixed price for local access with unlimited calling, and per-minute charges for interexchange calling, though local costs are not all fixed, and long-distance costs are not all variable.

Perceived value

Customers will pay more for a service that provides more value to them. When the service in question is telecommunications, customers are hard pressed to disclose what that value is. Some customers who complain about the price of local service pay a multiple of that amount, per month, for electric power or cable television. Customers who were willing to pay a certain percentage of their income for telephone service decades ago pay much less now, according to recent studies. The consumer price index for telephone services has risen much more slowly than it has for services in general in the decades prior to divestiture and the time since divestiture. See Figure 11.1. Yet consumers are very sensitive to the prospect of paying more for telecommunications services and consider the low cost to be a right. There is little evidence that consumers are willing to acknowledge the low growth in price and concede the enhanced value of the local services they receive.

The lesson to be drawn for telecommunications marketers is that consumers have historically balked at pricing plans that appear to move the price of local service toward its cost, no matter what other benefits are included. This makes the argument for bundling services in such a way that prices for individual elements of service are not specified.

AT&T Wireless created a pricing structure to accomplish that. Wireless users found it difficult to manage—or even enumerate—their usage costs, because wireless rates often included a base charge, a per-minute charge for service, toll charges, and roaming charges for calls originated outside of the home area. A traveling customer could have, for example, a roaming charge, an airtime charge, and a toll charge on a single call. All of these charges are larger, of course, than their wireline counterparts. Customers who are willing to pay the higher price for convenience are often unhappy with the unpredictability of their monthly bills. AT&T's

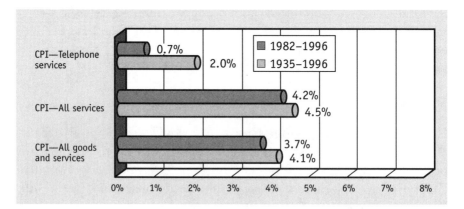

Figure 11.1   Relative growth in telecommunications services prices. (*Source:* FCC.)

Digital One Rate program was the first to eliminate roaming charges and long-distance charges for wireless users in many areas. The per-minute charge is equal for local and for long-distance calls and is slightly higher than an equivalent charge for a wireline long-distance call. AT&T's strategy is to equalize wireline and wireless service as much as possible to limit its requirement to build outside plant facilities to serve customers.

Pricing desires other than low price

Customers are drawn to price factors other than low price. Some customers would prefer to control their spending and are receptive to fixed-price packages. Such packages can be priced profitably; some customers would exchange a higher average price for the ability to budget the same amount every month. Flat rates are convenient, and customers do not wonder if each use of the service is worth its unit cost.

Other customers prefer usage-based pricing. Either these customers want to control their level of expenditure, or they are not sure how often they will use the product and do not want to commit. Customers who charge expenses back to employers or clients want to maintain an accurate record of usage.

IXC LCI created a price differentiation factor for itself when it began to advertise a 6-second, versus its competitors' 1-minute, billing measurement. Price-sensitive callers, its targeted market, recognized the value of paying for smaller time measurements. Whether LCI managed to cut its costs in offering this program is debatable, but it provided a recognizable brand and competitive advantage until its competitors met its standard.

Cost basis

In the long run, and across all services, costs must be covered by prices. Services provided below cost will be over-consumed, and services priced too high above cost will not sell if substitutes are available.

Long-term pricing strategies are better if they price products above *economic cost*.

Economic cost refers to the actual cost of providing a product, increased by a rate of return. The actual cost of the product includes the following:

- The cost to purchase the goods or their raw materials;

- Transportation;

- Manufacturing, if any;

- Marketing;

- Distribution and sales;

- Administrative costs;

- Any other costs that can directly or indirectly be applied to the product.

The assumption is that the price has to exceed what a simple investment in an interest-bearing instrument would have earned, accounting for investment risks in each situation. For the purposes of pricing theory, this just means that the price needs to exceed cost by more than all out-of-pocket costs combined, and accountants can do the rest of the math.

Arbitrage occurs when one seller can purchase a product at a very low price and then sell it at an especially high price. In markets undergoing deregulation, arbitrage happens frequently, but it does not last long. Arbitrage distorts markets and benefits and harms market participants in ways that are not deserved or sustainable. When the long-distance market was deregulated, new market entrants resold long-distance services in only the most lucrative markets, those that provided subsidies to higher cost markets. Arbitrageurs were not required or committed to serve all users as the larger carriers were. The resellers could earn higher-than-normal profits until regulatory and market forces normalized.

International callback is another example of arbitrage. National regulatory authorities set the prices for international calls originating within the country. Because each country sets its own rates according to its needs, prices for calls between two countries can vary greatly depending on where they originate. Country A sets the prices for calls that terminate in country B, and country B sets the price for calls in the other direction. This phenomenon has created an arbitrage opportunity that has resulted in the callback market. Country A charges $1 for a call to country B. Country B charges $5 for a call in the other direction. An entrepreneur offers calls at $3.50 to citizens of country B, then intercepts the call and changes the direction of the connection. For the $1 charge for the call from country A and the cost of reversing the connection (plus marketing and economic costs), the callback operator can earn a large profit until the markets equalize.

The emerging arbitrage opportunity is Internet-based telephony. This enables providers to offer interexchange, international, and fax services over the Internet backbone for prices that are significantly lower than their switched-network alternatives. Internet providers in the United States are now shielded from certain charges that are applied to Internet-based calling. Therefore, they can offer similar services, using the same local and interexchange facilities, at a very low price.

Arbitrage often prods markets to improve their structures. In the long-distance market, arbitrage added to the pressure on regulators and providers to reduce the actual costs and the regulated subsidy costs of providing service. International arbitrage is influencing countries with very high international subsidies to move actively toward privatization or deregulation of telecommunications. IP telephony is considered by many to represent the most likely future architecture for telecommunications. If arbitrage moves the market toward a more efficient or technologically sound alternative, then it has positive effect.

Provider objectives

Companies sometimes set prices to meet their corporate strategies. Occasionally prices are set that do not cover costs or that provide an exceptionally high profit. Low prices can be used as a way to gain market share. Companies sometimes provide a new product on a promotional basis to customers in the expectation that the customer will be willing to pay its full price later. Local exchange companies frequently offer central office-based services, such as call waiting for a month or two without charge, in the hope that the customer will enjoy it. Online services often send their software with promotional offers to sign new subscribers.

Loss leaders are products that are priced intentionally below cost to entice customers to purchase other, higher-margin products. Sprint has used loss leaders to sign subscribers to its long-distance service. For small business, it offered free calling for one day a week. This strategy is safer with businesses than with consumers. Sprint understood that most business long-distance calling is nondiscretionary. No sensible businessperson will wait until Friday to make a call that needs to be made today. Thus, Sprint was able to entice customers to its near-commodity service by offering a benefit that was appreciated but would not be abused.

Similarly, Sprint offered a marketing tie-in with Monday night football television programming, where residential customers could receive up to 500 free minutes per month between 7 P.M. and 11 P.M. on Monday nights during football season. A sponsor of the National Football League, Sprint earns brand recognition either way—by its banners and commercials on television and its free calling while away from the television. Providing the free calls during the game also ensures that the network is not strained; serious football fans let nothing interrupt their viewing. Other family members are placated and perhaps lured out of the room.

A company with a service that has a strong market and no competitors can offer its service at a high price, well above cost. One reason the local exchange companies offer their enhanced services using promotions is that the price of these services exceeds its costs substantially. Until competitors offer substitutes at a low price, the premium services will continue to sell and provide a high profit margin.

Competitive alternatives

Internet access provides the clearest model for what pricing might have been if telecommunications had been invented in 1990. When they first appeared, online services without Internet access (i.e., AOL, CompuServe, and Prodigy) offered a variety of plans for users. Most plans provided some hours at a base charge and additional hours for more. Promotions normally included software to get started and perhaps a month of free service to climb the learning curve. Consequently, the price structure was based on class (low-end versus frequent user) and usage beyond the allocated hours. In some cases, the customers who agreed to a higher base level of service were charged less for additional hours. In the telecommunications industry, wireless plans often share this characteristic. Part of the reasoning for this is to encourage online customers to commit to a higher number of hours per month, which provides a higher guaranteed revenue stream.

In the beginning, most online services offered only proprietary content, analogous to local telephone service without long-distance access. The user connected to the ISP server accessed only limited content and communicated with only the ISP's members. The introduction of the Internet to online services changed the customer base, the providers, and the price structure. Once the Internet was the primary driver of customers, rather than local content, many new providers emerged.

The online services (AOL and its competitors) offered expensive local content and many access points. Other services (ISPs) offered access points but little or no proprietary content, or no content and a single access point. In any case, their costs were much lower, and they seized the opportunity to change the price structure to something cheaper and simpler. As soon as the ISPs offered unlimited Internet access for about $20 per month, there was pressure on the online services to provide a competitive offering [1].

AOL offered a similar price structure and nearly ruined its own business when it combined the price change with a marketing campaign. Freed from the hourly timer, users were increasing their online presence substantially. The network was not prepared to handle the new traffic. Facing irate customers and the threat of class action suits, AOL refrained from signing up new customers until its facilities were upgraded and the service problems were resolved.

The pricing of Internet access services is critical for several reasons. First, it represents what might have been a market-based pricing system for telecommunications services had regulation and older technologies not created the price structures that are so common now.

Second and more important, Internet access providers represent a formidable strategic competitor to cable companies and local and long-distance providers. The Internet companies have created a customer base almost entirely through marketing. The current infrastructure-based providers of communications services do not have a body of marketing successes yet.

Last, American law treats Internet providers differently on a cost basis from the local exchange companies. Soon, the two types of providers are likely to meet in a battle for IP-based telephony. At present, ISPs do not participate in paying subsidies and some taxes. Without a change in the law, this can provide ISPs with a cost advantage that could translate into a price advantage.

Ancillary costs

While most telecommunications services simply involve sales fulfill-ment, some record keeping, and use, others require more installation or investment activity and create one-time charges for customers. These additional charges, or ancillary costs, can vary significantly among com-peting vendors. High-volume customers, mostly businesses, often con-duct analyses to capture the one-time and ongoing costs of several alternatives and compare them on an equitable basis. Consumers are often irritated by extra costs, sometimes because the initial charges are onerous, and sometimes because it complicates comparing alternatives.

For example, cellular wireless services, before PCS was introduced, amortized the one-time charge for the telephone unit as part of the base monthly price. Cellular providers and customers had little choice, as the cost of the phone approached $1,000 in the early years. Customers were willing to pay a higher monthly rate to eliminate the front-end costs. To recover their investment in the phones, cellular providers needed to commit customers to payments over months or years, so the one- to two-year service contract became the standard relationship between the wire-less seller and buyer. Agreeably, the service contract insulated the cellu-lar provider from an uncertain revenue stream. Unfortunately, it also ensured that customers watched new subscribers' per-minute prices fall for the duration of the contract without a chance to switch, even with one's own carrier. The long-term commitment also discouraged custom-ers from trading up to new hardware until the contract expired.

ISDN struggled with ancillary costs in its effort to gain market share. Among other problems, ISDN purchasers faced ancillary costs for instal-lation, and frequently these buyers needed to purchase a specialized modem as well. While these costs often provide switching barriers for products that customers already have, they also serve as entry barriers that can prevent a customer from trying ISDN.

Anticipated future prices

Technology has created a customer expectation that prices tend to fall rather than rise for technology-driven services. This expectation is warranted as, for decades, technology has advanced well in excess of inflationary pressures. While this phenomenon is generally not true in other markets, computer purchasers and customers of usage-priced services such as long-distance and wireless expect prices to drop. In the United States, some of the drop in long-distance pricing is due to regulatory restructuring and some is due to productivity improvement. It only matters, nevertheless, that customers see prices fall and behave accordingly. Customers who can delay the purchase of a product will do so in the hope that the product will improve, and the price will fall.

This is one reason that PCS providers needed to eliminate the long-term contract in their attempts to entice customers from cellular providers. Cellular customers tied themselves to carriers for a year or more and watched per-minute rates drop during the entire contract. For the most part, the long contracts were necessary for funding the investment in the telephone set, but they proved to be one of many pricing differences that differentiated and sold the PCS service to buyers.

Stickiness

Prices tend to be *sticky*, that is, prices tend to stay the same even when their underlying elements change. When a provider's costs for raw materials decrease, the provider can choose to leave the price alone temporarily to gain more profit until competitors lower their prices. After all, the customer was already willing to pay the higher price before the costs went down. Lowering prices involves some administrative cost. Raising prices a short time after lowering them erases the goodwill that the price drop would have created and can in fact anger the customer. Companies most often find ways to leverage the cost advantage that do not involve changing the product price.

Sometimes companies keep the price the same but add a bonus product to generate customer loyalty. With a consumer product, that can occur in the form of a larger container of the product for the same price, or cross-marketing a different product. A telecommunications provider can choose to offer bonus long-distance minutes with an expiration date rather than reducing a subscription rate. In any case, companies do not want to lower prices until they are reasonably certain that the underlying low costs can be sustained.

Similarly, companies try not to raise prices, even in the face of cost pressures or other changes to price factors. Cost increases do not provide the flexibility of reductions. Companies attempt instead to find substitutes for the elements that increased in price or reduce other cost components to compensate.

Value of bundled services

If the previous considerations do not provide enough direction in pricing, bundling services is a separate issue. Customers have stated often that they prefer to purchase their telecommunications services from a single provider. Providers prefer that customers purchase a package of services as well. (Because the concept of bundling is so important to pricing, it is covered in a separate chapter. Bundling characteristics and strategies are discussed in Chapter 12.)

Rate structures

While the needs listed above must be addressed, companies still have alternatives in recovering their costs. *Variable costs* are costs that rise with an increase in usage. *Fixed costs* are present no matter what the usage will be. Once a provider has decided to serve a geographical area, there are facilities required and fixed costs whether one customer or hundreds buy service. Commissions paid to sales representatives, as a percentage of revenues, are variable: The higher the revenue, the higher the cost.

Rate structures, all other things being equal, should reflect the underlying costs. Nevertheless, all other things are never equal, including—or especially—the preferences of customers.

Software, as one example, has an enormous fixed cost (the costs of development and distribution) and a very small variable cost (the price of the storage media or download activity). Software developers need to have an accurate understanding of the number of copies that they will sell before they can decide on pricing for a package. Software directed toward a small niche market needs to be more expensive than a similar product that appeals to most computer users.

Moreover, virtually all costs can be made more fixed or more variable. Hiring staff adds to fixed costs; outsourcing on an hourly or piecework basis creates variable cost. Similar actions are available for equipment and other services. When fixed costs are high, the product will be unprofitable until high volumes are achieved, and then profits are higher than they would be when variable costs are high. As a rule, when all costs are variable, all volumes are equally profitable.

When AOL and other service providers introduced *flat-rate* per-month pricing, none were prepared for the additional network traffic that it would bring. Some went back to a measured or modified flat rate, and others, like AOL, kept the flat rate structure but increased the subscription rate. Many of the ISPs discovered that flat rates draw subscribers. Because so many of the costs of operating a network are fixed, the flat rates added to profits, even when traffic exceeded expectations. The online providers that survived became stronger.

Providers enjoy some benefits with flat-rate pricing. Billing for long-distance services becomes much easier when providers move to a per-minute rate that is neither time-of-day nor distance-sensitive. Modifying rate tables in information systems becomes a manageable task rather than a momentous one. Administering usage-based pricing can comprise as much as 18% of the total cost of service [2]. There is a value to companies when they can predict the level of revenues because customers pay simpler, flatter rates.

Flat-rate pricing has its risks, especially when networks have a significant variable cost component. Users feel freer to increase their consumption when they do not worry about the charge for each use. The provider gets no additional revenue for the added service. For Internet usage, knowledgeable users begin to use more bandwidth as well, downloading images, software, and audio. Web hosting sites notice the new traffic and enhance their sites, adding pages and more bandwidth-consuming graphics. Eventually, the service provider needs to increase the network investment or risk degrading network performance and angering, or losing, customers. The additional investment will require either an increase in the subscription rate, as AOL was forced to do, or a reduction in profit, or both.

Telecommunications providers that view flat-rate charges as strategically necessary need to coordinate this objective with other corporate divisions. Because pricing needs to match costs to the degree feasible, companies that commit to a flat pricing scheme should try to reduce variable costs in favor of fixed costs. All departments need to work together to ensure that products will sell and that the business is profitable. In a monopoly environment, where all prices combined need to cover all costs combined, pricing for an individual service is not as critical. In a competitive market, customers can buy only unprofitable services from one provider and all other services from its competitors. This, of course, is unsustainable for the provider that is stuck with the unprofitable service. As competition intensifies, and companies have more control over their pricing, it is more important that marketers appreciate the revenue and cost components of the business.

Measured or usage-based services have advantages, too, for the provider and the customer. Customers believe that they have more control over their costs when they can control their usage. Providers benefit because most customers will at least hang up or log off when they know they will not be using the service, and they know that they will be charged for it, no matter what the amount.

Telecommunications providers are experimenting with alternatives to minute-by-minute pricing. The six-second (or one-second) standard unit may replace the one-minute unit, if prices stabilize. (If prices drop substantially, too many calls would involve fractions of pennies, and the too-small unit would become meaningless.) Some data services use packet size as the unit of measure for pricing. For services that are very expensive, such as satellite communications, small measures are valued by both customers and providers. Sprint was the first long-distance provider to offer a service without per-minute charges at all. In 1998, Sprint launched a $25 per month price for all weekend usage for its subscribers. The wisdom of this strategy is that it competes through low prices to gain high-volume subscribers and minimizes the cost impact by funneling traffic to low-usage periods.

As a rule, it appears that at least consumers prefer flat-rate prices to measured prices. The common perception is that usage-based pricing benefits the provider, not the consumer.

When two customers pay different amounts for essentially the same product, *price discrimination* exists. Price discrimination also occurs when the same price is applied to two different products. Coffee with cream and sugar always has the same price as black coffee, for example. Most price discrimination, though, concerns two prices for the same product. Some price discrimination, specifically price discrimination that lessens competition, is prohibited by the Robinson-Patman Act. Most price discrimination is legal, though.

Historically, monopolists such as telecommunications providers practice price discrimination with ease. Business lines were always priced higher than residential lines. Ameritech has filed to equalize its business and residential rates and eliminate that particular form of price discrimination. Price discrimination is desirable in many situations and very common. Movie theaters provide discounts to seniors or in the daytime to utilize their resources most productively. Software upgrades for existing customers—or customers of competitors—cost less than the same software for new customers. Airlines have created a science out of pricing virtually identical seats.

While it seems exasperating to pay $600 for a ticket and sit next to someone who paid $200 for virtually the same product, price discrimination can benefit both the high- and the low-price customer. Airlines sell the seats that they probably would not sell at the higher price, and those seats are beyond the break-even point, providing mostly profit. These seats are sold at low prices with difficult conditions for the customer to meet, such as a Saturday-night stay and a long lead time.

The vacationer who is quite willing to meet these onerous conditions benefits, as the low price puts discretionary travel within one's reach. The airline obtains additional revenues at a very low incremental cost. The business traveler will profit if the airline, faced with substantial competition for the business traveler, reduces its full fare through the profitability of the additional passengers. The addition of the low-price vacationer can also increase the number of flights available to the business traveler.

Price discrimination only works if the provider can identify the classes of customers who are willing to pay more, and then separate the customers. The separation is necessary to prevent customers who would have been willing to pay a high price from buying the low-price product. Airlines have a Saturday-night stay restriction, which eliminates the business traveler. Telephone companies used to price long-distance calls based on time-of-day and set low rates after 10 P.M. and on weekends. This ensures that business users pay the highest prices. Wireless providers offer free weekend calling for the same reason. Companies need to separate markets to practice price discrimination. If they fail to do so, they risk losing their high-profit customers to arbitrage.

Price discrimination will be an essential ingredient in telecommunications pricing, once markets are segmented and regulatory requirements are either eliminated or well-understood.

Pricing requirements

Customers have demonstrated that they appreciate simplicity in pricing. Providers prefer straightforward pricing structures, too. They enjoy operational savings when measurement is uncomplicated or unnecessary. They have access to more billing options and fewer billing costs. They reduce the rate of billing inquiries, which can be substantial and can be causes or symptoms of customer frustration.

Prices must be sustainable. A company that sells a product or service well below its actual cost will eventually be unprofitable, unless it can practice superior price discrimination. Unlike a monopoly, competitors need to earn back their investment and variable costs for every product in the long term.

Customers need to believe that they control their purchases. Companies will undoubtedly offer several bundles of services to customers. Bundling services can give customers the simplicity they desire, while an array of packages can offer the customization of services that provides the control they need.

Promotions and price wars

In general use, most promotions involve price discounts or bonus services for new customers who purchase services. While discounts are commonly offered to customers through telemarketing, so far, telecommunications providers have not taken outstanding initiatives in offering promotions. For long-distance services, few promotions are noteworthy, and they are nearly indistinguishable from one carrier to the next. Some examples follow:

- Long-distance providers often offer low per-minute rates to new customers in exchange for presubscription to their services.

- The promotion frequently offers reimbursement of the local company's cost to switch carriers or a refund for services at the level of the switching charge.

- Some long-distance providers have offered a very low rate for a short period, such as six months, or free minutes to be used in the first few months.

- Often, providers offer a check to new subscribers. Signing the check converts the customer and provides an audit trail against slamming.

These are unexceptional marketing techniques, and they have not reduced the level of churn significantly in the industry. The first offer, of a low per-minute price, does not qualify as a promotion in the strictest sense, as it usually lasts for the length of the contract. If they are based on a reduction of costs for the industry, competitors can offer the same package at any time. If not, the provider has found a genuine competitive advantage.

Offering to cover the local company's charge to switch carriers is not much of an enticement to the customer to switch. First, many customers properly perceive the fee to change carriers to be an administrative charge to the long-distance provider. Second, the customer can avoid the charge simply by staying with the existing provider.

Providers that offer a very low rate for a short time can be offering the short-term savings as a loss leader to gain the customer. This is likely to backfire; when the promotional period is over, the customer will switch to the least expensive offer that emerges. It is possible, though risky, for the loss leader strategy to offer competitive advantage. This can occur if the provider is certain that its low price will cover its costs after the promotional period (as prices drop), and if the provider believes that the promotional rate will still be competitive after the period. In a market-place as volatile as communications, these are dangerous assumptions. Offering a cash rebate for subscribing constitutes a significant marketing cost and one that draws a disproportionate amount of customers with a propensity to churn.

Price wars, while common, generally benefit only the customer. Beyond the obvious impact on the provider's profits, price wars have subtler and more dangerous long-range effects. Customers suspect that falling prices signal that prices were inflated in the first place. Price wars place too much emphasis on price and downplay other buying considerations. Price wars confuse customers. Promotional pricing for new customers annoys loyal customers who are not eligible for the discounts.

SELF-ASSESSMENT—PRICING

These questions will assist telecommunications marketers in pricing services in a competitive telecommunications marketplace.

- What are your costs underlying the product or service to be priced? Are they primarily fixed or variable? How flexible is the cost structure if pricing requirements are inflexible?

- How does your cost structure compare to those of your competitors?

- Do you have any products or brand equity that competitors simply cannot match?

- What are the rate preferences of your target market(s)? Would they prefer usage-based pricing or flat pricing?

- What strategic objectives will be met by pricing uniquely?

- What can you do to simplify your price structure?

References

[1] Bushaus, Dawn, "Flat-Out Fiasco," *tele.com,* Vol. 2, No. 4, p. 94.

[2] Gadecki, Cathy, "The Price Point, *tele.com,* Vol. 2, No. 12, pp. 56–58, 60, 62.

12

Bundling

Definition

Bundling is the packaging of different telecommunications services for customers. Usually, telecommunications bundling simply refers to local and long-distance services, but it is also used to describe the combination of either or both of those services with Internet access, cable television, wireless, and any other service broadly described as telecommunications. In the United States, customers want bundling and so do carriers.

Bundling services is not the same as promotional pricing. In a promotion, a provider might offer one service at full-price and a second at half-price for a limited time. Eventually, the preferential pricing of a promotion ends. A bundle of services is offered for the long term. If discounts are applied, they last for the duration of the subscription. Promotions are short-term customer incentives; bundling constitutes products.

Occasionally, the term convergence is used to mean bundling. Convergence refers to the trend under which the differences among a variety of technologies are disappearing. Cable companies are providing voice and Internet access services over the coaxial cable in their networks. Consumers are using their home voice lines to access the Internet, and soon the Internet is likely to support a large part of long-distance and international telecommunications, including fax. High-density television broadcasts will be digital, presenting opportunities to merge services. Convergence, then, represents the trend under which formerly separate services are merging, and bundling will represent the packages of service that result from convergence and other trends.

Customer desires

In a 1997 J. D. Power and Associates survey of more than 6,100 U.S. households, about two-thirds of respondents indicated that they would likely switch all of their services to one company [1]. Most respondents said that receiving one bill was a reason to bundle with one company.

The Strategis Group has found in its 1998 survey that nearly 85% of mid-sized firms are interested in buying bundled telecommunications services from one provider. Almost half of firms that are likely to buy bundled services would like the package to include local, long-distance, wireless, Internet, and enhanced data services [2]. A similar study concluded that nearly 80% of U.S. households would like some combination of local and long-distance services, cable television, cellular, paging, and Internet access from a single provider [3].

Of course, many of these endorsements are hypothetical, as bundled basic telecommunications services are not widely available or well-known by potential customers. In general, customers who lean toward bundling in surveys undoubtedly have an impression about what services and features would be included and what the price will be. These impressions will be tested against what providers eventually offer.

Examples of bundling

Wireless providers have bundled services as a competitive differentiation tactic. Cellular providers bundled airtime minutes into their service offerings to serve as a competitive weapon against their local competitors. PCS providers, eager to attract the customers of cellular providers, have leveraged the digital capabilities of the PCS technology to bundle news and stock quotes, messaging services, and other enhanced services in the basic price of the wireless service. Paging services frequently offer stock quotes and voice mail in their subscription prices.

Local exchange carriers have discovered a synergy between their high-speed digital subscriber line (DSL) offerings and Internet access. IXCs have found markets by bundling long-distance and other services such as Internet access. MCI offers its long-distance subscribers a lower-than-market price for Internet access. In another example of bundling, in the United Kingdom, cable operator Nynex Cablecomms Ltd. (now part of Cable & Wireless) offered its customers free local calling within its franchised area if they would subscribe to two premium-priced cable television channels. Nonetheless, bundling of local exchange and long-distance services is the most anticipated form of bundling, and it is not yet widely available to consumers.

Demographic packaging has become popular. Telecommunications providers offer packages tailored to specific demographic groups, such as families, or couples with no children. The United Kingdom's Bell Cablemedia (now also part of Cable & Wireless) and BellSouth in the United States have developed product bundles to reach certain demographic groups.

While hardware is not often mentioned as a candidate for bundling, there are examples of including hardware with telecommunications services subscriptions and including telecommunications services with hardware. Wireless providers often include most or all of the cost of the mobile phone in the monthly contract rate. Eliminating initial costs makes wireless service more affordable, although it does add to the monthly rate. Decades ago, residential customers rented all of the hardware, including the wiring and the telephone set. Thus, including the cost of the hardware in the monthly service charge as a way to sell services has a long history.

Bundling services into hardware purchases is a much newer selling tactic. Certain purchases of Gateway computers, which include Gateway's proprietary Internet access, provide for users a trade-up option that includes Internet access. Gateway has targeted the home computer user, who is often a first-time Internet user, and has offered a bundle of services that eliminates much of the purchase risk and allays the fears held by the buyers.

Benefits of bundling

Bundling offers benefits to both customers and providers. Customers want a single bill for telecommunications services. Until the AT&T divestiture in 1984, all customers had a single provider and a single bill for local and long-distance service. Customers served by non-Bell companies in the United States still receive one, if they want to. Customers appreciate the convenience of a single bill, and they hope that the combined services represented on the bill will offer them a lower price than services selected from a variety of carriers.

Customers also like simplicity, and bundled services offer simplicity. Many customers would prefer to bundle whether or not the aggregate bill was lower than their current costs for services. Bundling can also offer one-stop shopping and convenience for the customer. Ideally, customers who purchase bundled services also have only one source for customer support. This requires the provider to streamline its own organization and business processes, but at least bundling makes this opportunity possible.

Providers love the idea of bundling. For those providers that are currently prohibited from entering certain markets (LECs for long distance and IXCs for local service), bundling represents the opportunity to enter new markets without requiring large investments in new facilities—wires and brick-and-mortar. Bundling will require smaller investments in existing facilities.

Bundling creates an excellent cross-selling opportunity. It is many times easier to sell a new product to an existing customer than to sell existing products to a new customer. The more links to a customer, the more opportunities there are to sell more products and services. More importantly, the more links to a customer, the fewer ties from the customer to competitors.

Bundling services also generates customer loyalty. When a customer uses the same carrier for local, long-distance, and other services, the customer is unlikely to be coaxed to switch to a new long-distance carrier.

The bundle itself might not be available from any competitors. This resilience can be magnified if the bundled package the customer already has offers discounts for the incremental services. The new pricing or bills would constitute a significant barrier to switching for the customer.

To be sure, bundling local service with long-distance is not the most creative marketing technique that can be visualized by telecommunications providers. Bundling of these services adds new (or formerly prohibited) opportunities to an existing portfolio. More accurately, this represents a rebundling of services that were unbundled by regulators to create competitive markets. For the most part, local and long-distance bundling is the way to undo the fragmentation of the telecommunications industry caused by deregulating it one piece at a time.

Fortunately for buyers, the industry convergence that creates bundling opportunities also creates a fiercely competitive marketplace. Only the largest wireline, wireless, interexchange, and cable providers can afford to develop the infrastructure to support bundling. They will probably sell bundling with the same intensity of competition they demonstrate in the long-distance market.

On the plus side, bundling offers an opportunity for providers to differentiate their services. Many wireless providers use packages to attract subscribers to their own service, although the basic wireless offering is largely a commodity. Agents can package minutes in a unique way or offer free minutes at certain times. For the customer looking for a certain exact package, bundling provides what the customer wants at the provider's profitable price.

Bundling also consolidates customer information for customer service or additional sales. Through bundling, and with adequate information systems support, any statement of account gives a most accurate view of the customer's relationship with any division of the company that is represented in the package.

The economics of bundling

Telecommunications providers should be able to offer some form of discount to customers who buy packages of services instead of service elements. Many LECs already offer reduced incremental prices for additional central office-based services for a single line. Local exchange companies have developed creative pricing for second lines into the home.

Telecommunications providers can utilize bundling to reduce effective prices while maintaining an overall price structure. Wireless prices dropped by nearly 50% in the 10 years beginning in 1988. Providers have steadily raised the number of bundled minutes in their monthly plans, especially for the heaviest users. As shown in Figure 12.1, in only three years, the number of bundled minutes increased from 40% to 230%. In this manner, wireless providers can compete effectively, pass along lower costs to their customers, and maintain revenues at the previous level.

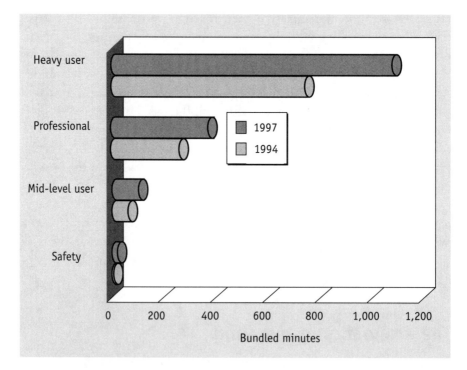

Figure 12.1   Rise in bundled wireless minutes, 1994–1997. (*Source:* The Strategis Group.)

Providers also benefit when bundling creates a stronger bond with the customer. The customer who already purchases three monthly services will be more receptive to a fourth than would be the receiver of a cold call. Serving one customer with four lines is less expensive than serving four single-line customers. Network efficiencies can be gained, because investment for growth is easier than new investment. Network usage created by new bundling customers presents higher network utilization and potential economies of scale.

While the U.S. telecommunications industry evolves toward deregulation, remnants of regulation still encourage certain actions by the industry participants. CLECs can buy certain elements of the local network from the incumbent providers, then rebundle them the way they were and sell them at a lower price. Local exchange companies have been encouraged by regulatory rulings to restructure certain types of traffic to prevent them being rebundled and resold. Not all bundling and unbundling will be driven by customer desires.

Bundling can offer economic advantages to the customer beyond any potential discounts. For example, as services become more complex and feature-laden, overlap between vendor services can occur. Even now, customers can purchase long-distance services from the LEC and similar Internet-based services from their ISP. As long distance is a measured service, customers would not mind the overlap. Nevertheless, they would be annoyed if each long-distance option carried a monthly cover charge. This economic cost would be avoided if services were bundled by a single provider.

On the other hand, consumers do not always make decisions based on their inherent economies. They have not flocked to the lowest price services that are available to them and might not migrate to bundled services in the anticipated rush. An estimated 75% of Americans do not subscribe to a calling plan [4], for example, although many of those households would save money if they did.

Some customers do not want to depend on a single provider for all their services. They will be willing to conduct the analysis necessary to find the least expensive communications components available from separate providers, which could indeed cost less than a bundled service. This customer needs to have whatever skills would be necessary to integrate the offerings of a variety of low-cost providers.

The bundled bill

Billing for bundled services simplifies the process for both the customer and the provider. In addition to the service package, the bill itself has bundling and unbundling opportunities. Different customers will want the bill structured in various ways. Some large customers that charge users for services will want the bill divided by organizational division. The ability of the telecommunications service provider to meet the information needs of customers will attract customers and discourage them from managing the telecommunications infrastructure in-house.

Other customers want to report their charges by type of service. Some families will want separate bills for their household and children's telephones. The role of technology in billing is discussed in Chapter 14.

The challenges of bundling

Bundling offers many benefits to customers and providers, but it includes several challenges as well. Most of today's telecommunications providers have a limited number of ready services they can offer to the market. Some are limited by regulation, some by geography, and others by their present investment in infrastructure. Many of the largest carriers have stated strongly that they intend to offer a full array of services, and some already have names for the offerings. The services themselves, nevertheless, are not yet available to the general customer base.

The fact that the services are not yet developed or permitted is not the only challenge. With the relaxation of regulatory restrictions and an enormous amount of capital, any of these giants will be able to enter any additional business they choose. The challenge is that virtually none of these providers has been active in these newly accessible markets for decades, and perhaps never has been. The drive to bundle services will force companies to offer services that they have no experience in selling, building, or supporting and whose technology is changing rapidly. Bundling services, even services that are well-understood, involves substantial change to business processes. Bundling services that contain their own mysteries will be more difficult.

The J. D. Power study also found that bundlers render their own unique characteristics, challenges, and additional costs. While bundlers spend about 30% more than others do on long-distance calling, they also call for customer service more often and are more active in the selection of service providers. They are twice as likely to switch carriers. There is a clear correlation between customer satisfaction and their willingness to bundle. To get customers to buy service bundles, a provider will need to make them happy and that, too, costs money.

A third challenge is to develop simple service packages that meet the needs of the widest array of customers without creating confusion. The permutations of five or six types of service, such as local, high-speed data, long-distance, cable, Internet, and enhanced services are staggering. Add parameters from each service to the mix—such as "two custom calling services" or "sixty minutes per month"—and the level of complexity grows quickly. The value of bundling is partly to reduce marketplace clutter, not add to it.

Business customers will be eager to bundle the same services that consumers desire, and more. AT&T has made overtures toward billing and bundling the basic telecommunications services and adding 800 service, teleconferencing, virtual networks, domestic and international private line, and frame relay services, among others. Large business customers can handle more complexity than consumers, but the business segment is often more eager to conduct a vendor comparison of exactly equivalent services. Too many or too few options can eliminate a provider from consideration.

Designing packages needs to begin with the requirements of the customers likely to bundle. The number of bundled packages to be created cannot be unlimited, but most customers should find a package that comes close to their expectations. Alternatively, some companies will consider bundles in which services can be substituted at the customer's discretion (any four services for a fixed monthly price). Obviously, any services offered in this manner must be roughly the same cost to the provider. BT offers bundles of service that include up to 250 minutes of local weekend calling and a choice of up to six enhanced services, resulting in a monthly discount approaching 50%. Because these residential services are still offered in an environment not yet fully deregulated, the prices can be discounted substantially and remain profitable. In a truly competitive and unbundled arena, it will be more challenging for competitors to price services attractively in bundles.

Another challenge could be caused by barriers in the organizational structure of the provider. Some telecommunications providers already offer several of the services that can be bundled successfully, but they have not yet bundled them. Some delays can be due to regulatory restrictions, such as the need to use a separate subsidiary to offer competitive services. Once the restrictions are lifted, massive organizational change will be required before services can be bundled, especially if they are priced favorably. Divisions will have to agree on their own portion of the bundled price, and joint customer care is necessary. Funding any advertising, infrastructure costs, or promotional pricing would need to be negotiated by multiple divisions, each of whose profitability is measured separately.

Moreover, bundling changes the skills required of sales personnel, billing and service representatives, and virtually any employee who has customer contact. Each employee needs to be familiar with the broad product line offered by the company because customers should be able to resolve their needs with a single point of contact.

Separate subsidiary requirements and other regulatory restrictions have prevented certain types of bundling in the United States and the European Union. While services are regulated, any marketing tactic that strengthens a monopoly or near-monopoly carrier is cause for regulatory concern. Regulatory compliance can also lead to legacy pricing expectations by customers, organizational anomalies, and other circumstances that are unfavorable to successful bundling.

Bundling services will require that providers continually stay up-to-date about customer needs and competitive offerings. Customers who want to bundle services can change their minds as to what services they would like to bundle. An A.T. Kearney study compared what customers wanted in a bundle in November 1996 and again in June 1998 [5]. Long-distance and local services were ranked highest in both surveys. Wireless services and paging increased significantly, but a cable service option in a package decreased in popularity. Providers that bundle service will need to conduct market research persistently to monitor the offerings of competitors, identify new customer needs, and manage the life cycle of existing bundled products.

The case against bundling

Some customers will not be candidates for bundled services. Generally, the customers with the most specific needs, the most in-house talent, or the largest telecommunications requirement are unlikely to seek out bundled services. Companies will always have to provide service components for customers to coordinate to their own needs.

Research has shown that customers are not favorably disposed toward all possible bundles. RKS Research & Consulting conducted a study that demonstrated that residential customers would resist packages that include telephone, cable, and electric services [6]. Moreover, the most affluent and educated customers were the least inclined to purchase bundled services.

Customers with unique and inflexible requirements will not find a service package to meet their needs. Other customers have very simple needs or limited buying power. For those customers, the extra cost, if there is one, to customize service and purchase only what they require, is well-spent.

While bundling can generate customer loyalty, some customers regard the service bundle as a leash. Suppose a customer joins an IXC because the bundled Internet service and e-mail address is priced attractively. Later, a competitor offers the package components, separately or combined, at a lower price. The customer will find it difficult to leave because the e-mail address acts as a barrier to switching. Certain customers will balk at packages that limit their future options.

Some customers, particularly large enterprise networks, already support an in-house telecommunications staff. Networks of this size rarely find packaged solutions for their needs, for the same reasons that they need an in-house staff to meet proprietary requirements.

Telecommunications services often need to fit into existing and planned facilities owned by the company. This is also true of the largest customers. These network operators are more likely to go to the same suppliers as the telecommunications carriers and establish their own network.

Bundling dissimilar services can add the headaches of one partner to the revenues of the other. It is not as straightforward to deny cellular service for nonpayment when local regulations protect the local service portion of the package. At the very least, terminating one element of the package can create pricing problems for the rest of the bundle.

Still, bundling will meet the needs of many consumers and some businesses. Most telecommunications providers will undoubtedly offer packages as an option, rather than a mandate.

SELF-ASSESSMENT—BUNDLING

To bundle services successfully, telecommunications marketers require answers to the following questions.

- Will your targeted customer segments respond to bundling?

- Which services will your customers want to bundle?

- Do you currently offer services that could be bundled?

- What cost reductions can you gain through bundling that you can pass to customers?

- Can you include services in the bundles that act as switching barriers?

- Would customers be interested in the bundles you can provide?

- Are there products that are obviously missing from the bundles you can offer? Can you partner with other providers or acquire companies to fill out the product line?

- Is your organizational structure prepared to support bundled services? Is the billing system flexible? Are the customer contact groups prepared to expand their knowledge and their communications with other groups?

References

[1] Nanji, Zaiba, and Kirk Parsons, "So Many Choices," *Telephony*, Vol. 233, No. 2, p. 34.

[2] Strategis Group, "Business Branding and Bundling Telecommunications Services: 1998," 1998.

[3] Meade, Peter, "Is Bundling Really Better?", *America's Network*, Vol. 100, No. 18, p. 26.

[4] Britt, Phillip J., "No Easy Money," *Telephony*, Vol. 234, No. 23, pp. 232–236.

[5] Schmelling, Sarah, "Bundling Takes On New Meaning," *Telephony*, Vol. 235, No. 2, pp. 20–28.

[6] Morri, Aldo, "Bungling on Bundling," *Telephony*, Vol. 233, No. 13, p. 14.

13

Advertising

Advertising strategy

Advertising does not create sales. Instead, advertising predisposes customers toward buying a product or service by changing or reinforcing their attitudes. Several conditions must be present for the potential customer to consider buying: The customer must want a particular benefit, and the customer must believe that the service will offer that benefit. Advertising can address those conditions by providing needed information to the customer. For the customer to complete the transaction, the perceived value of the service must exceed its price, and the distribution channel to get the service to the customer must be in operation. Advertising cannot complete the sale. This does not imply that advertising is less than crucial in the telecommunications market. It is necessary but simply not sufficient.

Telecommunications providers, like other marketers, have learned that advertising strategy is a corporate initiative, not a departmental one.

Successful creative programs last for years and present a unified image for the provider in all of its target markets. Each brand presentation, segment tactic, and individual program needs to support the overall marketing strategy.

Most professional advertisers maintain their own processes for creating advertising strategies, but they typically contain several elements in common. One component is a definition of the *target audience*. This is generally one of the first tasks in the strategy, as its resolution will drive many of the other activities. Another important component is the *advertising objective*. The objective often relates to the company's *market positioning*, which originates in the corporate strategy. The thought process surrounding the objective considers the *desired response* from the target audience. These responses can take the form of belief or attitude change, corporate recognition, or purchases of one's own service instead of a competitor's. A strategy statement usually contains *support* for the decision on behalf of the buyer, including the benefits the buyer will gain.

Once the overall strategy is set, the telecommunications provider, usually with the assistance of an advertising agency, develops the advertising program. A creative strategy for an incumbent local service provider targeted to the consumer market will be very different from that of a facilities wholesaler, as shown by Table 13.1.

Not every local provider will have the same objectives or develop the same strategy, and local providers that go through the planning process

Table 13.1
Components of Advertising Strategy, by Market

	Consumer Local Service	Wholesale Networks
Target audience	Consumers	Managers of large networks, resellers
Objectives	Retain market share	Optimize network utilization
Market positioning	Leadership	Top three wholesalers
Desired response	Brand awareness	Lead generation
Benefits promised	Good service at competitive prices	Network reliability

with the same objectives can reach very different conclusions about their most effective approach. Next, *media planning* involves selecting the proper media channels to gain exposure for the advertised services. This process first selects among media alternatives such as broadcast and print. Other tasks include choosing the geographical coverage for the campaign and developing an appropriate and affordable schedule. The resulting presentation to the marketplace, in the form of advertising, community activities that strengthen the advertising message, and marketing materials, should be consistent with the decisions made in the creative strategy process.

Changing factors in telecommunications advertising

Advertising is not a new requirement of deregulation. Competitive markets are voracious consumers of advertising, but advertising is important to monopolies as well. Advertising for telecommunications services is already very different from the campaigns of the past, and it needs to be. The stakes are higher for participants in competitive markets.

One of the most successful campaigns in history is the AT&T "Reach Out and Touch Someone" program, which spanned many years of monopoly marketing and extended into the beginning of competition. The purpose of the campaign was to increase the overall usage of long-distance calling among consumers. AT&T recognized that long-distance calling was both expensive and highly discretionary for consumers, and its campaigns emphasized the heartwarming benefits of voice contact.

The "Reach Out and Touch Someone" campaign focused on the emotional benefits offered by long-distance calling. By doing so, AT&T chose not to emphasize other potential benefits of calling as compared to its alternatives, such as avoiding the cost of travel or the effort involved in writing a personal letter. The assumption underlying the campaign was that a logical rationale for long-distance calling would be less effective than a sentimental approach.

The important element of differentiating the service from its immediate competitors was unnecessary in a monopoly environment. Monopolists need to expand the market, not market share. AT&T's goal was to increase consumer calling, which would provide network utilization during the lower usage periods of weeknights and weekends. Promotional pricing helped to steer the traffic toward the desired periods. The incremental calling was profitable at the discounted rate, which was still well above actual economic cost.

Even in a competitive environment, a market leader often uses a similar strategy to increase the overall consumption of the services it sells without a direct effort to increase share. Kodak's "capture the moment" and Nike's "just do it" campaigns illustrate this point. This approach, when successful, results in a larger market, and provides the leader with a proportionate share of new customers. AT&T continued to promote market growth rather than product differentiation well after competitors began to gain share. On the other hand, it was necessary for competitors MCI and Sprint to differentiate their services in ways that would attract new customers. MCI chose to differentiate by emphasizing its low price, and Sprint, in its successful "pin drop" campaign, worked to differentiate on quality of service. Ironically, MCI and Sprint prices were closer to each other's than to AT&T, and the near-commodity long-distance services were unlikely to serve as a differentiating factor for either of these competitors or AT&T.

Deregulation has also resulted in consolidation of advertising programs. This trend is a consequence of company mergers, the imminent recombining of local and long-distance services, and corporate decisions to coordinate a single advertising strategy. An associated trend to commit the advertising function to a single advertising agency is partly because of the overall consolidation that is taking place in the telecommunications industry. This decision is supported by the agencies' added consultative and strategic responsibilities in developing a comprehensive strategy. An added advantage of concentrating an advertising account with one agency is that larger accounts enable agencies to develop the industry and service expertise to develop more creative and effective programs.

A major change for telecommunications providers is the amount of investment required by the hotly competitive market. Telecommunications changed from a low-profile, relatively unchanging industry to among the top three advertisers [1]. AT&T's long-distance sales and advertising expenses rose from $50 million to $800 million in the decade following divestiture [2].

Advertising expenditures for the industry leaders grew 30% in the first quarter of 1997 over their levels a year earlier [3]. AT&T, MCI, and Sprint ranked in the top 15 brands in the first half of 1997, according to *Advertising Age* [4]. Advertising expenditures will continue to rise as new competitive markets emerge, even while prices are driven downward by competitive pressures. To introduce its flat-rate "Digital One Rate" wireless service, AT&T launched a $35 million television, radio, and print campaign. AT&T's elimination of roaming and long-distance charges positions the service as a wireless alternative to wireline local service, and the campaign was intended to ensure that the strategy was communicated to customers.

The rise in advertising will force telecommunications providers to target their advertising as they do the rest of their businesses. Both Sprint and AT&T have developed branding programs targeted at specific markets in addition to, and consistent with, their corporate creative strategies [5].

Strategies in the telecommunications markets

MCI launched its 1-800-COLLECT service with a successful campaign supported by advertising and direct mail. The company then looked for other services that would use the same inexpensive sales channels. Dial-around services (then reached by five-digit codes such as 10-XXX, now requiring a seven-digit prefix) were gaining acceptance among cost-conscious consumers. MCI's 1990 Telecom USA acquisition represented MCI's entry into the dial-around business. MCI was aware that most dial-around providers were small and unable to respond competitively to a major advertising effort. The successful 1997 national rollout gave MCI market leadership in dial-around, although MCI's corporate name was never used in the advertisements years after the acquisition was completed. The success of Telecom USA provided a boost for the hundreds of otherwise unknown providers by giving visibility to dial-around services. In 1998, AT&T leveraged the service recognition and responded with its own Lucky Dog dial-around service. Like the MCI offering, AT&T's service scrupulously avoids any prominent reference to the parent company. In both cases, the services are positioned as very low cost. It would not enhance the branding of either service to associate them with the market leaders.

Sprint created a local PCS radio campaign in California that surpassed its expectations in both awareness and sales effectiveness [6]. Two characters intended to introduce the target market to Sprint PCS service became cultural icons in the area, earning recognition by talk radio and competitive advertising. The message and the humor of the ad were that the two characters could continue to hold an inane conversation because the service was affordable. Sprint's risk-taking humor, coupled with its market selection, resulted in a campaign that lasted months longer than the original plan and gained valuable brand recognition for the provider.

Other wireless carriers have used different techniques to humanize their services. Both PrimeCo (a consortium of U.S. carriers partnering to provide PCS service) and Sprint use animated characters as mascots for the service. Omnipoint Communications utilizes a parrot asking for a blind date in a continental inflection over the wireless phone.

Sprint also runs a series of television advertisements that emphasize the clarity of the digital connection. Each version shows a caller on a wireless phone who makes a connection and stammers a convoluted sentence. In one version, a frightened suitor asks a father for the daughter's hand. In another, a harried executive issues obfuscated orders to his stockbroker. In a third version, a mechanic tries to explain to a customer what is wrong with a car. In each case, the caller is handed a Sprint PCS phone and magically makes a single accurate statement using the PCS service. While not entirely logical (since both phones cannot be on the same connection at the same time), the ads are both entertaining and informative. In all of the above examples, the wireless operator attempts to defuse the primary consumer concerns about wireless service: cost, connections, and technology.

AirTouch Cellular launched a Spanish-language campaign directed to the Los Angeles Latino market [7]. Media included Spanish television and radio ads, Spanish print ads, billboards, and sales literature. The strategy includes Spanish-language back-office functions, including customer service representatives and voice-mail prompts. All of the three leading IXCs have launched similar campaigns for targeting customers for long-distance service.

Telecommunications providers have been less successful with positioning their unified network products than they have been with branding single service offerings. MCI's "Gramercy Press" campaign was intended to introduce the networkMCI service. The campaign was creative but generally not successful. The television commercials were intended to work as short situation-comedy vignettes. The companion Web site, like the commercials, was innovative and entertaining but did not focus on the specifics of the branded product. Last, the product itself was poorly defined, so the campaign underscored its vagueness.

The more recent MCI campaign "Is this a great time, or what? :)" is more successful if its goal is overall brand recognition and expansion of the market, and not as successful if its goal is to gain market share. This is a strategy for a market leader to adopt. It is less applicable to a company with less than a third of the interexchange market and a much smaller portion of the huge telecommunications services market. MCI, nevertheless, has always enjoyed more brand recognition than was warranted by its share, due in part to its highly advanced marketing. The emoticon (a combination of typed symbols to express a feeling) included in the slogan engenders a sense of Internet savvy in the campaign and positions MCI as a leader in the Web-enabled marketplace.

Advertising and the courtroom

Telecommunications providers sustained a long tradition of fighting competitive battles in the courtroom, and advertising is no exception. Much of this sparring occurs between IXCs and local wireline service providers. MCI's Ivan campaign was apparently intended to increase the pressure to deregulate local service, comparing the lack of marketplace choice in the United States to Russia under communism. Local carriers responded with angry press releases and spoofs.

AT&T and Bell Atlantic underwent a press battle when AT&T advertised a competitive directory assistance service. Bell Atlantic also sued AT&T for advertising savings compared to Bell Atlantic's rates on certain toll calls. In addition, large facilities-based providers have sued dial-around carriers for misleading advertising.

GTE filed a lawsuit against AT&T to stop an advertising campaign that singled out its service. AT&T's campaign was launched when GTE introduced long-distance service in certain areas served by AT&T. The ads claimed that GTE's network was not as reliable as AT&T's for certain categories of calling.

Voluntary self-regulation in the advertising industry has brought forth several organizations, including the National Advertising Review Board (NARB). MCI challenged AT&T's implication in its advertising that the company supports the most powerful Internet network. After several decisions against it, AT&T agreed to change its ads in deference to the NARB.

IXCs and local wireline carriers, on their own and through their trade associations, continuously post press releases and advertising intended for customers, defending their positions on controversial reforms in the path toward deregulation. In addition to print and television advertising, unbranded Web sites present the views of industry participants.

One of the chief reasons that long-distance rate advertising has been misleading is that the rates were simply too complicated to communicate through traditional advertising. Some long-distance rates, including dial-around, offer savings after a certain number of minutes per call. Some services have a monthly charge or a setup charge. Others promise savings over AT&T's highest nondiscounted rates. Ideally, as rate plans become more straightforward, the rate of litigation will decrease.

Whether the rush to the courtroom or the arbitrator is an automatic reaction from formerly regulated companies remains unknown. Most business issues for the leading telecommunications providers, including MCI, have historically been resolved through the regulatory process. Deregulation will eventually force these competitors to resolve their differences more amicably (to interconnect), more expeditiously (to meet market demands), and more courteously (as long as customers judge their providers on image as well as value).

Budgeting for advertising

Budgeting is an important component of the planning process. Advertising in mass markets is expensive, and it is difficult to measure its direct results. Funds not allocated to advertising can be redirected to product promotion, campaigns to retain customers, other marketing programs, or nonmarketing uses.

One of the alternative uses for the resources budgeted to advertising is promoting the product either through a temporary reduction in prices at the point-of-sale or offering incentives to distribution channels. As a rule of thumb, when customers are unaware of the services, but distribution channels are strong, advertising is more effective than promotion. When the reverse is true, if distributors are weak but customer acceptance is strong, promotions are more effective.

Managers use several methods for budgeting advertising dollars, some more complex than others. A common method is the *percentage-of-sales* method. This technique allocates a certain percentage of company revenues to the advertising budget, then adjusts the percentage over time. A flaw in this method is that it reverses the actual cause-and-effect relationship between advertising and sales. Advertising creates sales, and its expenditure should not be driven by their historical levels. In a marketplace undergoing deregulation, this is an especially dangerous method. For most providers, those newly entering the market and those losing share, the revenue level changes too quickly month-to-month for the percentage to have meaning. For incumbent providers whose advertising budget was based on a monopoly market, the advertising budget needs to increase significantly and quickly, or market share will drop precipitously.

The *competitive parity* method of budgeting for sales will undoubtedly be popular among competitive telecommunications providers. Telecommunications providers have always used benchmarks to measure their operating efficiency, growth, and other success factors against those of their peers. The operating companies divested from AT&T were fiercely competitive siblings, especially when they were all about the same size. This method of developing an advertising budget can be achieved in two ways, through percentages of gross sales or dollar-for-dollar. In the *percentage approach*, companies evaluate their competitors' expenditures as a percentage of their sales, then they budget the same percentage of their own sales. In the *absolute approach*, the provider determines what the competitor spent, then matches the expenditure dollar-for-dollar. One criticism of this method is that it presumes that one's competitors are more knowledgeable than one's own bottom-up or top-down approach would be. In addition, the method fails to take into account differences in overall strategy.

Table 13.2 summarizes 1996 advertising expenditures for the top four long-distance providers.

Note, in Table 13.2, the wide disparity between MCI's and World-Com's premerger advertising expenditures as a proportion of revenues. Table 13.2 suggests that MCI was practicing the absolute approach of competitive parity to the degree it could afford. If it matched only the percentage investment of its competitor AT&T, its message would be overpowered by the market leader. MCI's apparent goal is to be visible in a marketplace where it simply cannot afford AT&T's level of advertising investment. Sprint, too, is taking an absolute approach but against a different target. Sprint most likely views MCI as its nearest competitor. Its investment is almost the same as that of MCI's, but it represents nearly double the revenue percentage for the company. WorldCom's acquisition of MCI was undoubtedly influenced by its admiration of MCI's marketing skills, especially as compared with its own. In 1996, their market shares were disparate by a factor of four, but their marketing expenditures were geometrically opposed by a factor of 75. Some analysts fear that WorldCom, in an effort to gain scale economies, will reduce the company's advertising investment, thereby unraveling one of the benefits that justified the acquisition [9]. Advertising strategy is a philosophy, not a science.

Table 13.2
Relative Ad Spending for Long-Distance Providers (*After:* [8])

Rank	Provider	1996 Share	1996 Advertising ($M)	Ad Dollars Per Share Point ($000s)
1	AT&T	60.6%	$539.2	$ 8,897.7
2	MCI	22.3%	$297.1	$13,322.9
3	Sprint	11.1%	$245.2	$22,090.1
4	WorldCom	6.0%	$ 4.0	$ 666.7

The *task and objective* method for budgeting defines specific objectives, the tactics to achieve those objectives, and the cost of those tactics. This method is gaining in popularity, although it is the most difficult to execute. The task and objective method is also especially well-suited to new products, where historical and competitive data are not available.

One generally disdained method for setting an advertising budget is to allocate whatever funds are left after accounting for fixed and variable expenses and a desired profit. The *affordable* method is used with success only in situations in which the company is very small and funded poorly, or in desperate straits, needing new sales.

Measuring the effectiveness of advertising

As the price of advertising and the number of potential venues have multiplied, advertising costs have risen significantly. Companies investing significant amounts in advertising are always interested in knowing the value of their investment. Advertising effectiveness can be measured, but the value of the measurements has been a source of controversy.

The first issue is that advertising never occurs in a vacuum. Customers are always exposed to messages in the news, from competitive providers, and by word-of-mouth. The lag between the message or messages and the purchase increases the difficulty of associating a particular advertising tactic with aggregate sales, let alone a particular sale. Recreating the advertising environment with focus groups is arguably not comparable to an individual's receptivity to advertising in daily life.

The most problematic issue with measuring effectiveness is also the most controllable—setting measurable objectives in the first place. "Increase sales" is a weak advertising objective. The most practical benchmarks include a measurement technique (gross sales), an increment (by 4%), and other limits as appropriate (in the southwest region this year). Sales constitute a frequent measure, but other measures such as name recognition or migrating customers to more profitable services are also valid.

The measurement process often begins before the advertising campaign is launched. If the purpose of a campaign is name recognition, the precampaign analysis needs to assess the baseline measurement. In fact, a known baseline measure should be a part of the objective-setting process prior to the campaign design.

Companies test campaigns as a way to avoid needless expenses, evaluate alternative programs, and optimize their advertising investment. Tests to measure brand recognition or attitudes toward competing brands require specifically designed studies outside of normal business transactions. Other required data such as sales and market share can be measured within the normal course of business, as long as the data are collected properly before the analysis.

New advertising venues

Broadcast television has provided a window on the general population for decades. Its strength is its reach to millions of potential customers, and the high cost of advertising reflects this. Cable television, on the other hand, enables advertisers to target smaller segments of customers. This ability will be especially useful when telecommunications services become more specialized as the market matures and technology develops further.

The Internet has created an outstanding new venue for advertising telecommunications services. Its many features include its low cost and ability to target consumers and the new opportunities it affords for customers to receive product information and advertising upon their own request. Compared to other advertising venues, such as print advertising and broadcast television, advertising on the Web is relatively inexpensive. Its reach is presently limited to those with computers, modems, and Internet access, but the industry is growing at a rapid pace.

Targeting potential customers is easy and direct. Users who request a word, phrase, or group of concepts at a search site can look at advertisements for products that match the search terms entered.

AT&T has established partnerships with Web portals like Excite, Infoseek, and Yahoo, and MCI has partnered with Yahoo [10]. Each of these providers has cobranded online services with the portals with whom they have marketing partnerships. Telecommunications providers have purchased banner ads on selected Web sites to advertise their consumer long-distance services, and marketers of wholesale services place more targeted advertising on the Web sites of trade publications and business periodicals.

The effectiveness of Web advertising is widely debated, but it is clear that the Internet offers excellent tools for measuring users' exposure to banner ads and their subsequent follow-up activities. An Internet user who sees a product of interest can click on the product's hyperlink and visit a full-screen advertisement and additional links. Furthermore, a user can request that a provider deliver advertisements via e-mail. The original banner advertisement that draws the user is designated as *pull technology,* and advertising requested by the Internet user is known as *push technology.* Pull technology involves the user seeking some advertisement or information of interest and then downloading, viewing, or pulling the information to the screen. This form of advertising is unprecedented in predecessor print and broadcast media. Push technology is more like conventional media, in that the advertiser or information provider sends new information to the user, at specified intervals or when the site is updated. The novelty of push technology, though, is that the provider sends only information that the user requests. The user can subscribe and then halt the subscription when the information is no longer needed.

The first applications of push technology downloaded news headlines to the desktop and refreshed the information whenever the user returned to the Web. Banner advertising is included with the headlines and stories. The more relevant application of push technology occurs when the advertising is not an artifact of the desired information; it is the requested information. Users register with sites to receive updates about promotions, discount pricing, and specifications about new products. The advertising is the message. The provider is overjoyed to reach a receptive potential customer at a low cost, and the user is delighted to learn about services that meet his or her needs in a manner that is completely within his or her control.

Measuring Web effectiveness is still in its infancy, but an abundance of information is available to those who want to analyze it. Companies sponsoring the advertising can count the number of hits on their site and request demographic information about users who visit their home pages for more information.

230 Marketing Telecommunications Services

SELF-ASSESSMENT—ADVERTISING

The following questions will assist telecommunications marketers in analyzing their companies' advertising strategy and programs.

- Does your company support a comprehensive advertising strategy? Does the program have the involvement and support of senior management?

- Are the strategy's objectives clear and measurable?

- Do all of the communications initiatives, internal and external, support the objectives of the advertising strategy?

- Are your present advertising campaigns most suited for a market leader or a company that is trying to gain share?

- Are your present campaigns targeted to the customers you need to reach?

- Are your campaigns intended to raise company awareness or brand recognition for specific services?

- What steps have you taken to use the Internet effectively as an advertising channel?

References

[1] Cleland, Kim, "Surging Telecom Industry Seeks Super Shops," *Advertising Age*, Vol. 67, Issue 48, p. 27.

[2] Noll, A. Michael, "The Effectiveness of Long-Distance Competition: Who Really Benefits?," *Telecommunications*, Americas Edition, Tele-Perspectives, Vol. 31, No. 3, p. 28.

[3] Interpublic's McCann-Erickson WorldGroup, remarks of Robert J. Coen, senior vice president, director of forecasting, reprinted in *Advertising Industry Outlook*, *Interpublic Insider's Report*, June 1997.

[4] "Top 200 Brands: January–June 1997," *Advertising Age*, Vol. 68, Issue 44.

[5] Cleland, Kim, "Sprint Moving Beyond Product Ads," *Advertising Age*, Vol. 68, Issue 19, p. 18.

[6] Schmelling, Sarah, "Only in California," *Telephony*, Volume 235, No. 12, p. 91.

[7] Davis, Stephania H., "Want to be Persuasive? Try Spanish," *Telephony*, Vol. 233, No. 7, p. 62.

[8] "Top Long-Distance Phone Services," *Advertising Age*, Vol. 68, Issue 39.

[9] Snyder, Beth, "Marketing Savvy Tops WorldCom's Bid for MCI," *Advertising Age*, Vol. 68, Issue 40, p. 63.

[10] Prince, Paul, "Buddy System Could Spawn 'Net Gain'," *tele.com*, Vol. 3, No. 9, p. 28.

Part V

Customer Focus

14

Customer Care

Customer care defined

Customer care refers to all the efforts made by a telecommunications provider to ensure that customer expectations are met or surpassed. Most customer care activities concern the three most frequent interactions between the customer and the provider: using the service itself, problem solving, and billing. The concept of customer service has been a part of marketing theory for decades, but the term *customer care* arose during the organizational effectiveness boom of the last decade. While the distinctions between the two terms are subtle, *customer service* is often deemed a cost of doing business, while *customer care* is viewed as a strategic investment in nurturing the customer relationship. The ramifications are not so subtle. Companies do whatever they can to minimize costs, but they will spend aggressively on strategically important functions. Those that view customer service as a cost of doing business are making a statement about their positioning in the marketplace. Providing minimally acceptable customer service rather than full-fledged customer care is a valid strategic choice, but it is indeed a choice. The rest of the marketing effort will be consistent with that decision. Typically, service providers that compete on the lowest price will opt for the lowest cost customer service as well. Telecommunications providers that plan to compete elsewhere in the marketplace will want to support a competitive, if not superior, customer care function.

Service and support can provide the basis for competitive advantage. Deregulation will require service providers to sustain competitive advantage or disappear. In the wireless market, the existence of two licenses in each market resulted in much less competition than the presence of four or more carriers after the introduction of PCS. High-quality customer service is an important opportunity for these providers to build customer loyalty. Customers often include customer service as an important purchase criterion. Moreover, the customer service transaction is an inexpensive form of primary market research. Data gathering efforts during customer service transactions help the service provider to identify new customer needs and areas of service vulnerability.

Business transactions between the customer and the service provider, other than the customer's normal service usage, can occur in several forms. Each form has its own evaluation criteria. Transactions can be paper-based, so factors such as the layout, clarity, and timeliness of the bill will affect the customer's experience. Some customers prefer telephone interaction for service. Calls for services will lead the customer either to a service representative or to an automated service system. Customers who balk at automated systems will require an option to connect to a service representative. Increasingly, such transactions are Web-based. Customer satisfaction with Web-based service is influenced by the timelines of e-mail support, the usability and value of the Web site, and the customer's ability to save time by using the Internet rather than alternatives.

Most often, the transactions are with the service provider's employees. This often presents a cross-functional undertaking. The service representative taking the call will need to complete an order through operations, put a repair ticket in motion, or work with the accounting function to resolve a billing issue. Each component of the transaction needs to work well in the judgment of the customer. Service representatives who create good transactions while setting up repair visits can be undermined by technicians who are late or incompetent, and vice versa.

Customer care frequently crosses suppliers as well. When service bureaus handle customer transactions in the name of the service provider, they are entrusted with the guardianship of the customer relationship. Telecommunications providers choose to outsource the strategically important customer care function for several reasons. Outsourcing can cut costs or at least match costs to the volume of business activity. The contract with the service bureau often allows for per-call charging. Telecommunications providers pay only for completed calls and eliminate the potential fixed costs of an in-house system. Service bureaus are better at handling peaks and valleys of activity by assigning representatives to the tasks as required. Customer care is the mission-critical function of service bureaus, and service bureaus frequently have sophisticated operations and measurement systems to ensure high-quality performance.

In a deregulated environment characterized by retail, resale, and bundling, customer care transactions will cross suppliers beyond those outsourced by a single provider. Many providers will choose to work with other providers to create service packages that operate seamlessly to customers. In this situation, there is no room for finger pointing when services do not perform to the customer's satisfaction. Providers will need to integrate their operations systems in a single display and generate sufficient diagnostic information for a single service representative to resolve the customer's problem. The need to integrate seamlessly underscores the importance the customer care function will have in an unregulated market. Many customers will seek bundled products solely to simplify their customer service activities. Customer care, then, is an integral part of the product.

Lessons from the ISPs

A TeleChoice study of ISP customers and their preferences found that two of the top five scoring attributes of customer satisfaction were service-related [1]. Those ISPs rated highest for customer service were also rated highly overall. Seven characteristics of superior customer service were identified:

- *Customer service is a focus.* Rather than emerging as an afterthought to the core business functions, customer service is placed at the center of the business.

- *Staff training is essential.* Managers at Digex, a business-focused ISP, estimate that staff training can consume 10–20% of the staff member's time.

- *Customer service personnel must have technical expertise.* Functional specialization is required to meet the technical challenges posed by customers of the customer service staff.

- *Customer support is involved in service design.* Customer service personnel are the first to identify service needs, and they can help to design and test new offerings. Sometimes the customer support solution to an individual's problem can become a standard service offering. Customer service also helps to identify features that need redesign and those that should be eliminated.

- *Excellent customer service anticipates the needs of customers.* Superior service representatives clamor for operations and usage data to provide network information to customers. First-class management infrastructure also enables customer service staff to tell customers who call in that the problem is being resolved.

- *Customer support keeps customers informed.* Techniques to accomplish this objective include the following:

    I Notifying individual customers of relevant news after the customer's call is over;

    I Announcing outages and estimated repair intervals;

    I Posting updates about general customer issues on the site;

    I Providing access to operational data for customers to monitor the service operation without calling for help.

- *The ISP constantly solicits customer feedback.* The questions are specific and designed to measure the ISP's performance in the eyes of the customer. This feedback is especially important when markets are highly segmented.

Some of the study conclusions are counterintuitive. Numerous providers would balk at announcing bad news such as service interruptions, especially to customers who otherwise would not have been aware of them. Many providers are not inclined to solicit feedback, believing that most of it takes more time to dispatch and analyze than its ultimate value merits. This is especially true when markets are very segmented because so much more data are needed for the results to be statistically valid. Providers have historically been wary of providing real-time access to network operations data. Telecommunications providers will need to evaluate their customer care strategies and their corporate cultures to decide how proactive and candid they want their services to be.

Features of customer care systems

Customer care systems require information systems to integrate various sources of data about the customer. Telecommunications providers often support a variety of legacy applications for billing, provisioning, ordering, and account tracking. Customer care systems draw data from each of these systems and integrate the data in a format usable by a service representative.

Customer care systems can define procedures for the service representative, so that each customer's service experience is standard and effective. These procedures are developed over time to be the most effective and efficient. Disruptions stemming from employee turnover among the service representatives are minimized when the procedures can then be used by the employee's replacement. Sharing case data enables a variety of service representatives to handle follow-up customer calls without frustrating the caller. Customers do not need to wait in a queue for the representative who opened the case. Customers can receive service 24 hours a day, seven days a week.

According to Innovation Resources, the percentage of customer support employees doubled in the last six years, and customer support costs can be as high as 15% of total revenue [2]. Companies have realized that superior customer care retains customers. Customers have learned to expect excellent customer care from all but the lowest price providers. The circular nature of this interaction places pressure on providers to offer exceptional customer care and to continue to improve their quality as customers demand more and competitors catch up.

Achieving the essential return on customer care investment requires that three conditions are met: customer satisfaction, operational effectiveness, and cost effectiveness. A customer care system that meets all of these criteria will provide additional profit and growth; a system meeting only one or two of them will eventually have an adverse effect on profitability.

Some telecommunications providers have developed their customer care systems in-house, building on and integrating their other in-house applications. Developing systems in-house is appealing because the internal staff is familiar with the business processes, the available data, and the existing applications. On the other hand, in-house development is time-consuming and expensive.

Purchasing customer care systems from outside developers can be more cost-effective because the development costs are spread over all of the customers. These systems will require customization, and they will never meet the exact specifications of an in-house development. Furthermore, unless the vendor is asked to customize the system extensively, it will be difficult to develop systems that differentiate the service provider from its competitors, which are able to purchase the same software. On the other hand, one advantage of buying packaged software is that the software developer takes responsibility for upgrading the system as technology demands. Maintenance to meet the requirements of individual customers will constitute an added cost of either in-house or vendor development.

Why invest in customer care?

Not all telecommunications providers will choose to invest precious resources in customer care systems. The logic of making customer care a priority stems from finding an acceptable basis of competition. There are three major areas for a company to differentiate itself from its competitors: price, the product itself, and customer service. Examples of telecommunications providers that choose each method are ubiquitous. Virtually all dial-around companies differentiate on price. They offer no customer service at all and make no particular claims about the quality of their services.

The market leaders often use network quality to differentiate their offerings. Both Sprint and AT&T financed branding campaigns to demonstrate the quality of their services.

Local exchange providers appear to be setting up brands for the inevitable competition by differentiating through customer service to the consumer. BellSouth has used the customer service approach in its campaign, "Nobody knows a neighbor like a neighbor." Ameritech uses the slogan, "In a world of technology, people make the difference."

It is no surprise that local exchange companies will choose customer care as the path to differentiation. Differentiation through pricing is an unattractive long-term solution for the largest providers. Smaller providers (or unbranded subsidiaries of large providers) can pick their markets and compete on price only in the most profitable areas. Large providers that attempt to compete with low prices find themselves in price wars. Furthermore, incumbent local service providers will find competing on price to be difficult until they substantially restructure their costs.

Differentiating through quality-of-service will eventually emerge as a strategy in the local services market, as it has in the long-distance market, depending on the carrier. Incumbent LECs with a strong reputation will be able to use the approach of service quality, but only in their existing geographical territories. They will be encumbered in new markets with the strength of other incumbent carriers' brands and will experience lower brand recognition of their own. Upon their divestiture from AT&T, incumbent local Bell carriers were prohibited from manufacturing equipment. Unlike their long-distance counterparts, they do not directly control the manufacture of their networks' components. This makes product differentiation more difficult for them in the short term. Moreover, telecommunications service is a near-commodity in performance. The quality approach, taken on a national basis, will be an uphill battle as it has been in the long-distance market. However, some carriers, including today's long-distance providers, will rise to the challenge and succeed.

Customer care, then, becomes a new option for sustaining the interest of customers. Local exchange providers in their existing territories have the home court advantage. The sizable investment required for world-class customer care systems provides a barrier to entry for the smaller market entrants. Competing through customer care avoids having to maintain the lowest prices or the highest quality network, a dubious proposition in a near-commodity market.

Another reason that companies will emphasize customer care is the window on the customer that it offers to the provider. For subscriber-based services such as telecommunications, customer care, including the billing system, is the only interaction the provider has with the customer.

Not all telecommunications providers will choose to differentiate through customer service—and with good reason. A study of the wireless market found that price was by far the most important consideration for customers [3]. Furthermore, in a comparison of factors, including those deemed important by customers and those for which customers reported satisfaction, customer service was the only factor with a higher reported satisfaction than its importance. In other words, customers were receiving better customer service than they needed. One interpretation of the study is that customers are not yet ready to use other purchase criteria because the technology is still expensive compared to wireline alternatives. Furthermore, price is frequently overvalued as a buying criterion in an immature market such as wireless, and eventually customer service could prevail. Therefore, the wireless company that decides to leverage its customer service will at least have a good head start. PrimeCo stated when it entered the market that it intended to differentiate by using customer care [4]. Its tools include databases and back-office systems to let customers choose between automated care and human interaction. Rather than compare its customer care to other telecommunications providers, it benchmarks against "world-class customer service providers" in all industries.

Help desk applications

Help desks have come about because products and services that are based in technology have become more sophisticated and harder for customers to use. The traditional telecommunications technology was simple to use but featureless. Instructions to pick up the telephone and make a call were all that were needed. Technology promises to provide more network capabilities, but more complexity as well. Help desks are in the future of those telecommunications providers that want to offer more than basic voice services to their customers.

More than half the technical support calls to ISP customer care are related to installation, usability, or education [5]. Some ISPs leverage their customer care infrastructure to create a positive customer relationship even before the customer has a problem. Others integrate their customer care with their product development efforts.

Help desks are supported by software applications that queue the calls, route the callers based on caller ID or product information, and connect the caller to technical support personnel. The database match can route the caller to a staff member who speaks the caller's preferred language. Innovations used in help desks include on-hold announcements of the caller's anticipated wait, pop-up screens with the customer or case history when the caller is connected to a support person, and no-wait fax back alternatives for frequently asked questions. Help desks can provide a window on unmet customer needs. Users sometimes call to find out how to use features they expect to find or hope are available. When the existing service does not support those features, these calls offer feedback to service developers about new features that should be evaluated for development.

The customer support history generated by the help desk enables the provider to build a base of knowledge that can be used by support personnel and those customers who prefer not to talk to a technical support representative. Previous and frequent problems can be listed in a knowledge base. Support personnel can search the knowledge base while they are in a conversation with a caller.

Many companies post their knowledge bases on their Web sites, searchable directly by users. These provide two advantages: cost avoidance when the user does not call directly and better service to the many users who prefer to conduct their own research. Other customer self-help applications will include online documentation, downloads of new software for software-based services such as IP telephony, and support of user news groups where users can share problem-solving tools, receive company-sponsored or peer-to-peer solutions, or ask for new features.

Provisioning and trouble-handling applications

Another goal of customer care is to ensure that customers are still attended to after the sales phase is complete. Customers occasionally suffer a bout of post-purchase regret, and a sometimes-valid suspicion that the end of the sale is the end of the courtship. This transition requires that customer orders are fulfilled within the customer's range of expectations, that services are tested and approved for operation by the technical staff and the customer, and that trouble reporting and resolution is efficient and effective. Meeting time commitments is often more important to customers than speed of fulfillment. The customer who expects the order to be installed today is unhappy when it is installed tomorrow morning. The customer who was assured at the same time that the order would be completed by end-of-business tomorrow will be delighted with the morning turnaround.

One very significant outcome of the transition to deregulation is that the total telecommunications services package will regularly be offered through the efforts of many service providers. Customer care, on the other hand, needs to be provided at a single point of contact. Facilities-based wholesale carriers will want to and be required to provide their retailers with electronic access to network information about provisioning, order status, billing, operations, maintenance, and repair. Both participants in the wholesale/retail relationship will need to support customer care systems to provide timely and accurate information to their own customers. Other multivendor environments will require that linked customer care systems are sophisticated and reliable.

SNET's Repair Access Provisioning System ties together data from four systems and gives repair personnel access to maintenance records, customer information, and technical information [6]. Repair technicians can examine a line prior to arriving on site. The system is credited with a productivity increase of up to 20%.

Billing applications

The basic requirements for telecommunications billing are to collect, rate, and calculate the charges for calls, then send a bill to the customer and receive payment. Billing can operate as a simple accounting function, as it has for most of the industry's history. In its most native form, billing is essential. Loss of a single day's billing records can result in millions of dollars in vanished revenues.

Not all companies are provided with a recurring opportunity to inter-act with customers, but billing enables telecommunications service pro-viders to enjoy repeated communications. Granted, receiving a bill is not always the happiest customer experience, but billing can indeed act as a positive monthly experience for customers and providers.

Large service providers have the opportunity to improve their cus-tomer relationship through the strategic application of billing. Unfortu-nately, most billing systems currently in use by telecommunications providers have their roots in 20-year old technology. In the past, the criti-cal nature of these applications compelled providers to change them by building on their existing architecture, rather than risking a new struc-tural foundation. The resulting system is inefficient, difficult to maintain, lacking in documentation, and unable to utilize newer platforms and features.

Fundamental changes required by regulators, such as unbundling of network elements and local number portability, will cause providers to make massive changes. Upgrades in the variety of service offerings in an unregulated environment will also encourage service providers to make big changes to their billing systems. The mergers taking place across the industry are forcing providers to consolidate or replace the legacy billing applications currently in use. Some service providers opt to integrate existing applications. Others face up to the radical changes inherent in their situation, select one of the merger partners' systems, and convert the other partner's data. Some view the merger as an opportunity to move all data to a completely new platform. Many changes brought on by mergers or deregulation are nondiscretionary, so continuing to patch old technology is no longer an option. This will enable providers to make additional changes that are strategic in nature. New technologies, such as relational databases, enterprise resource planning (ERP) applications, and client-server designs will be necessary to gain the benefits of a strate-gic billing application.

While the existing large providers have the advantage of experience, entrants to the industry can start with new billing applications and use their flexibility as a differentiating factor. MCI used its billing application to support its Friends and Family brand and gained significant market share from its efforts. If the legacy systems maintained by the incumbents no longer provide economies of scale, they indeed become a disadvantage in the new marketplace when newcomers benefit from state-of-the-art billing technologies.

Billing has the potential to create customer goodwill and assist providers in targeting markets and creating desirable services. Knowing the usage patterns of individual customers will enable providers to offer them more cost-effective discount plans before competitors do. Providers can offer seasonal promotions or promotions targeted to match usage, such as discounted calling to frequently called numbers. Suppose a customer uses a fee-based service, such as directory assistance or call-return, several times in every month. The service provider can add a promotional offer to the bottom of the bill for "five for a fixed price" to simplify the customer's calling plan. Technology enables the provider to react to individual needs, like a hotel's concierge, to ensure that each customer receives unique treatment during the business relationship.

Billing can meet the needs of individual customers in ways that are unrelated to the service offering itself. Bills can be presented in the language of the customer's choosing. The billing cycle can adapt to the customer's payment needs. Customers who are concerned about the environment will appreciate carriers' efforts to use recycled paper or print on both sides of the page. Business customers can receive billing information by line or by organizational department or enhanced with an analysis of usage trends over time. Large customers can pay their bills using electronic transfers.

The bill can help to establish the customer's confidence in the service provider, especially when rate comparisons are confusing. Long-distance carriers sometimes show the customer the discounted rate in the bill and show the savings the customer has achieved. This helps to prevent churn when a new carrier comes along with promised discounts that are no better than the customer's present plan.

Over time, billing for traditional services will undoubtedly get easier. Per-minute pricing replaced the point-to-point rate tables and simplified billing for both customers and providers. As local service moves towards deregulation, it is most likely that national providers will offer averaged prices for local service instead of the jurisdiction-by-jurisdiction mosaic that is now in effect. Billing will become more capable of handling complexity, but at least basic service will become more straightforward.

Integrated applications

In the local service market, carriers are already setting a foundation for a strategy of differentiation through customer care. BellSouth, a three-year J. D. Power & Associates customer satisfaction winner, supports a customer care system integrating sales, service, and billing applications [7]. Customers can sign up for new services on the company's Web site. Among the benefits of customer self-service is that customers' input is sometimes more accurate when it does not involve keystrokes made by a service representative. Customers who prefer a personal interaction can also work through the company's call center.

Ameritech instituted an around-the-clock customer service center that allows customers to call with questions, service requests, or orders [8]. Establishing the center was part of the company's initiative to heighten brand recognition.

The Internet is an emerging venue for providing superior customer care. US West allows its customers to review summaries of the bills online on a secure site [9]. Its customers can review its product catalog and determine if new services are available to them by their central office. GTE offers billing summaries to its wholesale customers.

The Internet enables service providers to customize their care to target markets, such as small business. Ameritech customers can learn about new area codes and order products for home offices on the Web. The Web site also contains sample letters for notifying business contacts when a customer's area code changes.

Requirements for a customer care system

Customer care processes should have certain characteristics to meet the needs of the customer and the telecommunications provider.

- Customers should receive prompt and friendly attention from the service representative.

- The service call should be resolved in a single session with a single representative.

- The system should contain all the information that the representative needs to complete the transaction.

- The provider's procedural and organizational problems should be invisible to the customer.

- Staffing, scripting, and support should be linked to other marketing efforts, such as promotions and advertising, to ensure that service representatives are knowledgeable and processes can fulfill marketing promises and objectives.

- The service representative should be empowered to accomplish all normal transactional tasks without involving a supervisor.

- Metrics to evaluate the quality of customer care must be in the long-term best interest of the customer and the provider. Speed of call completion will not serve as an appropriate long-term metric.

Telecommunications service providers that are committed to world-class customer care will develop targets in the sometimes-competing categories of customer satisfaction, cost of customer care, and operational performance. In doing so, they can manage dozens of measures at the same time. Within each category, the service provider can monitor more detailed metrics such as speed of resolution, training costs, number of transactions closed in one contact, and other relevant measures. Measures are selected based upon their importance to the successful positioning of the telecommunications provider. For example, a provider with an in-house customer care program is more concerned with attrition than one that outsources its customer care. A provider that chooses a minimum customer care strategy will be more concerned with cost per transaction or with self-service transactions as a percentage of all customer care transactions.

Unless comparable benchmarking data are available, the provider will establish targets by developing relevant metrics, measuring current performance against them, and setting attainable targets, reasonable time frames, and sufficient budgets. The final task involves establishing and conducting a monitoring system and revising targets and business processes as warranted by results.

Customer care information systems should support these processes. In addition, they should have the following characteristics:

- The system should be easy for the service representative to learn and easy to use.

- The screen should display required information and hide all information that is not relevant to the task at hand.

- Supporting databases need to be complete and accurate.

- New systems need to integrate with existing applications.

- The customer care system should collect transactional data to supply to management to eliminate future similar problems, to supervise customer service representatives effectively, to identify marketing opportunities, or to collect customer satisfaction measures.

- The system must be flexible and scalable enough to meet the changing needs of the telecommunications provider.

The variety of requirements for a customer care system, coupled with the abundance of competitive systems and service bureaus in the marketplace, underscore the significance of the provider's customer care strategy.

SELF-ASSESSMENT—CUSTOMER CARE

Answers to the following questions will provide telecommunications marketers with insight into the value of the customer care provided by their company.

- Is customer care strategically important to the future success of your company?

- Is customer care part of the differentiation strategy?

- Do you presently support an effective customer care system?

- Has customer service been a significant contributor to customer retention or to your churn?

- Do your competitors support world-class customer care systems? Do you need to maintain parity with those competitors who do, or will you compete on some other basis?

- If you have chosen to compete outside of customer care, is your customer service function properly sized and funded?

- Does your present customer care system have the flexibility and the scalability to meet future requirements?

- Should your information system be developed and maintained internally or do outside vendors offer better alternatives?

References

[1]   Wetzel, Rebecca, "Top-Notch Customer Service Requires Change in Focus," *Interactive Week*, February 6, 1998.

[2]   Kirchner, Soren R., "Transform Your Help Desk Into a Profit Center," *Call Center Solutions*, Vol. 17, No. 1.

[3]   O'Shea, Dan, "Looking for a Little Satisfaction," *Telephony*, Vol. 231, No. 6, p. 15.

[4]   Meyers, Jason, "Carrier Profile: Past, Present, and Future," *Telephony*, PCS Supplement, Vol. 230, No. 20, supp. p. 32.

[5]   Machatzke, Jens, and Ross Sullivan, "Do You Have What It Takes to Be an Internet Service Provider?," *Telephony*, Vol. 233, No. 21.

[6]   Thyfault, Mary E., "Clearer Connection With Customers," *InformationWeek*, Issue 700, pp. 245–248.

[7] Thyfault, Mary E., Stuart J. Johnston, and Jeff Sweat, "The Service Imperative," *InformationWeek*, Issue 703, pp. 44–55.

[8] Britt, Phillip J., "Call Us Any Time," *Telephony*, Vol. 234, No. 4, p 36.

[9] Guy, Sandra, "Working the Web," *Telephony Supplement: Internet Edge*, May 5, 1997, pp. 30–35.

15

Customer Profiling and Data Management

The need for customer profiling

Managing information has become virtually as important as managing networks. This will intensify in a competitive world. Already, the largest telecommunications providers probably invest a comparable amount in building and maintaining information systems as they do in their networks.

More than in other industries, this generates a philosophical question: Do customer billing records constitute information about the business or are they the business itself? Is network usage information valuable data about the network or about the customers?

Telecommunications providers realize that their businesses depend on how they manage the data they can collect and what they can learn from them.

The data warehouse

The notion of data warehouses and data mining receives much attention as telecommunications providers prepare for the cutthroat competition that awaits them. This attention is warranted, as the data warehousing concept will undoubtedly change the marketing profile of the most successful telecommunications providers. Nevertheless, the favored terms *warehouse* and *mining* and sometimes *mart,* frequently used to describe the repositories for data or the act of searching the data, hold implications that undermine their value.

The data *warehouse*, for example, has inspired many companies to invest immediately in creating databases whose main characteristic is size. A warehouse is a place to store data, and the best knowledge to have about a warehouse is a list of what resides there. Unlike an actual warehouse, having the data is not the most important feature of a data warehouse. Its value is in viewing and arranging the data. Similarly, excavating these repositories is sometimes described as *mining*. Again, while the term evokes images of finding gold, it implies incorrectly that the value is somehow inside the data themselves. This is not exactly the case. The data certainly need to have a certain level of accuracy, but that is not their benefit. Even *mart*, which generally refers to a subset of a larger database, implies that the product is complete and that it is just a matter of buying it and consuming it. None of the above designations captures the research and analysis required of the telecommunications marketer in conducting database marketing.

To conform to the most popular terminology, this chapter will use the term *warehouse* to describe the collection, analysis, and use of customer data. Customer data include information about customers, information created by customers (such as calling activity), and information that helps a company understand its customers and their buying patterns.

Why have a warehouse?

Users of data warehouses give many reasons for their creation and use. Interest in warehouses was sparked when telecommunications providers began to notice the level of churn that accompanies competitive markets, such as the churn in wireless markets compared to the more stable (and less competitive) wireline industry. Former monopolists with no experience in handling churn are sensibly troubled by it. Data warehouses can help companies understand and manage churn. According to Insight Research Corporation, churn will decrease by a conservative estimate of 5% when data warehouses are used effectively [1].

The information can help to identify new markets to target and new products to offer. Analysis can sort out the desirable profitable customers from the customers a company would like to send off to competitors to diminish their bottom lines rather than its own. Examining customer data in creative ways can provide not only information that is descriptive of a company's existing long-term customers, but predictive information to help the company find many more.

The warehouse can support management decisions concerning the proper distribution channels for various products, sorting out superior versus average performances for distributors and internal sales staff, and evaluating the profitability of entire market segments or product lines. Data warehouse analysis can help to predict the size of the market for a new brand, packaging, or price structure.

The warehouse can compensate—through various views of data—for the organizational boundaries within a company that interfere with customer care. Suppose your customer has offices on the East Coast and the West Coast. Perhaps your organizational structure interacts with each office through a separate division. Furthermore, the East Coast office is one of the company's best customers, and the West Coast office has no outstanding characteristics. If the West Coast office falls behind on payment, or asks for special treatment, the local account manager may decide that it is easier to let the account go than to make a special effort to keep it. From a local perspective, in a vacuum, this would be the proper course of action. It has important consequences to the corporate office, though, that need to be part of any decision of the local account manager. A data warehouse can identify the vulnerabilities caused by inevitable organizational boundaries.

While the justification for a warehouse infrastructure can simply come from increased sales and customer loyalty, other benefits have come to companies that have used the data to solve corporate problems. IXCs have used warehouse data to identify fraudulent calls within minutes of their occurrence by analyzing calling patterns against today's calling. Ameritech used data from its warehouse to fend off a mul-timillion-dollar lawsuit by AT&T when it was able to demonstrate that the IXC's claims of overcharge were exaggerated. Local exchange pro-viders have used warehouse data in conjunction with network decision support systems to pinpoint geographical areas in which demand for cer-tain services is high. A database analysis can uncover a relationship in the location of diagnostic alarms in the network. This could signal a facilities problem that was not yet severe enough to be noticed by customers and not yet a cause of a network failure. While not directly a marketing dis-covery, the data warehouse can help to resolve product failures before they occur, averting a marketing crisis.

In the warehouse

The warehouse contains three main elements:

- Customer data and other information;

- Some hardware and operating software that capture and store the data in preparation for analysis;

- A set of analytical tools that arrange the data in new ways and enable the user to view the results.

Each of these elements has its challenges. Collecting the *customer data* is a complicated venture. The company needs to decide if it already has the data that it needs. It needs a plan to transport the data from their cur-rent residence into the warehouse and a plan to update the warehouse fre-quently as new data become available. It needs to test a sample of the existing data to understand their level of accuracy and improve the data integrity if it falls short.

When more than one data source is involved, someone needs to clean up the small differences in two sets of data that keep computers from con-necting records that should be connected. If data from one source state that the customer lives on 4 Street, and the other says that the same cus-tomer lives on 4th St, does the database connect the two records? Do dif-ferent data sources name the data elements in the same way? These differences must be reconciled in a transformation process between the origin of the data and the warehouse itself. The reconciliation has to happen when the database is first populated and again every time new data are written to the database.

When corporate mergers take place, how will the customer data be merged? This can be a problem for companies undergoing deregulation, even if they have not merged. Often, regulated companies are required to offer very different services in the various jurisdictions (generally states) they serve. This can create several types of data, stored in different formats, within the same company. Decisions about how to resolve differences without compromising conclusions will be necessary. That's just the collection part.

Because most of the data are coming from other systems, they need to be reviewed and "cleansed." Frequently, the warehouse includes additional information about how data from the various sources will be modified to make them consistent and qualified to be placed in the database. There is a layer of additional transformation rules or records, called *metadata,* or "data about the data," to describe how the information is transformed in preparation for the warehouse. All of this preparation activity represents about 70% of the effort involved [2].

The data are often cross-functional or interdepartmental. While the emphasis is most often on customer data, product information, network configuration, or other internal information can be incorporated into the warehouse. Data from disparate systems introduce technical challenges. There are simple differences, such as differences in departmental terminology. Management issues in controlling the use of the information can pose a larger challenge.

Not all data come from internal company systems. Demographic data can be compared against the existing customer base to determine areas of the best customer penetration. AT&T used this type of analysis to support a saturation advertising campaign on public transportation in one city neighborhood. The very specific targeting opportunity enabled AT&T to provide a significant advertising presence for international calling for much less expense than an ordinary advertising campaign.

Retailers use weather data to examine trends in buying patterns. While it is generally known that the weather influences shoppers' access to stores, it was discovered through this type of analysis that shoppers purchase different products on rainy days. Retailers noticed this peculiarity and enjoyed a distinct increase in sales when they moved certain products to prominent locations when the weather was overcast.

Data warehouses often inspire companies to collect data that they never considered important. Some companies have learned that buying patterns, not just calling patterns, change at different times of day. Many sales transactions do not currently store time-of-day data, or month-to-month trends, or mid-week versus Friday. While it might not be difficult to add that information to a warehouse, the planners need to evaluate which data elements are required and what operational information systems will need modification to collect the necessary data.

Hardware and software constitute critical considerations in the development and use of a data warehouse. The increase in price/performance of hardware has made it possible to create warehouses with masses of historical information. In theory, virtually every transaction for years can be reproduced in the warehouse. Telecommunications companies have better records of customer behavior than most industries, and until recently, few bothered to analyze it.

In theory, the more information that can be captured in the warehouse, the better the warehouse will be. Hardware is the least expensive it has ever been, but it is not free, and there is virtually no limit to what data elements the imagination can add to the database for analysis.

In practice, creating unneeded data parameters can increase the size and cost of the database exponentially, and slow the query transactions to an unusable level. When information systems efforts require expensive capital outlays and long development timelines, the pressure for the system to work well and provide results is considerable. It makes sense to install the most meaningful entries to the warehouse at first and add the less certain requirements after the database is in use. Warehouse planners need to strike a balance between including a cumbersome amount of detail or irrelevant data and providing enough information to draw meaningful conclusions.

Which software should be used to manage the data? Telecommunications providers need to conduct a make-buy analysis as they do for virtually every information system. Warehouses consume an enormous amount of storage space and processing capability. Whatever system is selected, it must meet requirements for the long term. In general, companies will have a more successful transition if they do not attempt to build the intelligent platform in-house. Dozens of reputable software companies have made significant commitments to developing warehouse products. Furthermore, several software developers have created specialized systems specifically directed to telecommunications providers.

Many successful implementations use an outside systems integrator with the specialized skills to install an enterprise warehouse. Still, many telecommunications providers prefer to take responsibility for the installation on their own. About 80% of service provider data warehouses are established in-house by making a business case and then purchasing the hardware and software [3].

Hardware and software for marketing warehouses will have different requirements from customary information systems. A typical online application for a company will show a new customer order. The input form is the same for all new orders, and the result updates the database of pending orders, or customers, or whatever. The accuracy of that particular order is critical, but its value is not dependent on the accuracy of the other pending orders. That is the way most typical information systems and databases work.

In a data warehouse, these functions are nearly reversed. Most of the database is relatively static, in that its records are not updated during transactions against it. Instead, the transactions are dynamic, as in "what are the defining characteristics of people who want to buy voice mail?" The transactions vary by user and probably change every time the database is accessed. The accuracy of the question, perhaps more than the data residing in the database, is critical to providing a valid response. Viewing tools provide the user with the ability to make database queries and present the data in a useful manner.

According to most research, the most important characteristics of the warehouse's computing environment are scalability (the ability to grow), system performance, and value.

Companies do not simply use warehouses to report summaries or filtering of the data that they have collected. In the past, marketing information provided a way for the astute analyst to ask the right question ("Do students make longer phone calls than others?") and receive verification (or contradiction) of the result. Note that an astute analyst is the most important component of the intelligence of the transaction.

The value of warehouses surges when the analysis includes *sophisticated tools* to predict the future behavior of customers. Instead of posing a specific question and receiving an answer, users can request that the system look at a series of variables and determine whether there are unexpected relationships in the data. While intelligent users will improve the speed and quality of the results, they are no longer the primary driving force of the analysis.

New analysis involves neural networks, predictive modeling, and other scientific tools. With the assistance of large computer processors and complicated statistical devices, companies can predict the likeliest customers to buy more services, to create uncollectibles or fraud, or to leave for competitors. All this analysis gives the company the opportunity to recover the customer or the revenue before it is too late. The company can create or advertise new services, initiate calls to customers who are at risk to leave, or promote different pricing plans for customers deemed to be receptive to them. U.S. Cellular determined that first-year contracts that ended at Christmas frequently churned in November and December [4]. The company's antichurn promotions stemmed defections during this period by 10%.

Examples of warehouse systems

Virtually every large U.S. telecommunications provider is building and using systems to analyze customer and operations data to improve their marketing results. Asian, European, and mid-sized North American providers have lagged behind the large U.S. companies in this area.

In all likelihood, corporate interest in data warehouses varies directly with the imminence of deregulated competition. When companies feel directly threatened, as do the largest carriers in the United States, they seek data warehouses for the competitive edge that they can present and for the window on customer behavior that they provide. Among telecommunications providers with data warehouses in place, many of the large companies appear to be pleased with the results they are receiving. Few have categorically stated that the returns are positive so far, but most of the warehouses are still under construction.

Today, much of the data in a corporate warehouse is created for some other business purpose: network utilization data for network management, call measurement data for billing, and sales data for compensation. Until now, there has not been an obvious corporate need or ability to view how this data interacts. With significant computing power available at a low cost, companies can now discover trends they could never see in the past. They can discern that the top sales performing agent is also the one with the largest customer turnover, weakening profitability and signifying flawed customer screening. Perhaps the most orders for premium services come from neighborhoods with the most network failures. What if the most profitable customers are assigned the lowest rated customer service representatives? These facts will emerge only through creative discovery within a data warehouse.

Figure 15.1 depicts the typical flow of data into a warehouse and the possible outcomes of the data analysis. The warehouse can provide predictive assistance in finding potential market share gains and potential losses, with enough time to act decisively. Management will choose to develop new products to fulfill unmet needs, develop new strategies to address competitor actions, or strengthen existing brands to retain or capture customers. Results and conclusions from the warehouse will assist corporate functions other than marketing, so that process is recognized on Figure 15.1's chart as well. However, marketing is the reason for being of the customer warehouse. Gains in marketing must justify the effort and cost of maintaining it.

MCI began building its warehouse in 1993 with the intention of creating a relationship marketing system. Its repository is among the largest of any carrier, partly because of its characteristic commitment to strategic technology, and partly because of the sheer size of the company.

US West has used its warehouse to prescreen customers to avoid the cost of sending marketing materials to customers who are unlikely to buy its products. The company also discovered through its data modeling process that by offering training or free usage, it could reduce its losses from customers canceling new services. Similarly, Bell Canada credits its implementation of a data warehouse in 1996 for substantially increasing Internet subscriptions and producing a long prospect list of qualified leads.

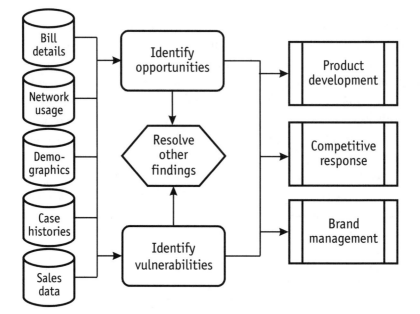

Figure 15.1   Warehouse data flows and processes.

360° Communications uses its warehouse to cross-sell cellular and long-distance services, perform credit analysis, and analyze churn to reduce it. Meanwhile, Ameritech has created a warehouse for its Customer Usage Tracking Systems program, and Bell Atlantic has utilized its data analysis to create a clearer image of its small business customers. Comcast Cellular, which serves about 800,000 subscribers, used its data warehouse to increase its direct marketing success rate by 20%. At the same time, the company doubled the number of new promotions it offers [5].

In fact, although small businesses are too diverse to make many assumptions about their characteristics, needs, and purchasing behavior, they are among the ideal targets of data analysis. Small businesses are well-suited as data analysis subjects for the following reasons:

- They are less price-sensitive than consumers and more loyal.

- They are receptive to profitable value-added services for which they can perceive a benefit.

- There are simply too many of them to justify too much investment in face-to-face sales.

Warehouses can assist telecommunications providers in finding new markets, new prospects, and new market segments. They can assist providers in retaining their existing customers by helping them to choose additional services or by noticing telltale signs that they are considering other competitors' offerings.

Types of analysis

Some of the analysis techniques used in data warehouses resemble those applied for market segmentation. These tools, which are based on the most sophisticated, computer-based artificial intelligence available, usually include the *classifiers, clustering, association,* and *sequence* techniques.

The *classifiers technique* takes available examples where the outcome is known and develops a model to apply to a larger population. The best known examples of this technique in the telecommunications industry are the fraud prevention efforts of IXCs. When their systems detect an uncharacteristic pattern of calling, such as international calls, many calls in a short period, or a single calling card used for simultaneous calls, credit is restricted.

Clustering looks for closely related records to use as a starting point for further analysis. In telecommunications, a cluster can include seasonality associated with the purchase of wireless products, demographic characteristics of groups who respond to offers in bill inserts, or groups with a similar payment history. These customer groups will provide a valuable basis for developing a segmentation strategy.

Nothing prevents a telecommunications provider from including a given customer in a variety of customer segments. A customer will be included within a geographical segment for one analysis, an industry group for another, and a group representing "customers of two years or more who don't have voice mail" for yet another analysis. With enough data, the segments can be so well defined that a customer will share five or 10 characteristics with the other members of the cluster.

Association tools find products that are frequently purchased together, such as beer and pretzels or cellular service and cellular accessories. Packages that include both services can be priced attractively enough to provide a significant increase in profit without a commensurate cost. More commonly, the warehouse identifies couplings that are counterintuitive, such as two very different products in a retail setting. Merchandisers can then take the products and place them in close proximity, assuming that buyers for one product will notice and purchase the other.

Sequence techniques identify products that are typically purchased in sequence. A sequence analysis would ascertain that customers who order voice-messaging service return within months to add voice-forwarding service. With that information, it would be valuable to follow up on voice-mail customers after a short time to offer them a promotional use of a forwarding service. A single customer can be tracked to establish whether calling patterns are changing over time. If that customer would benefit from switching to a different rate plan, it is important to let the customer know before the competition can. The overture can serve as a way to retain the customer and generate loyalty. The provider that behaves proactively will retain the customer at a slightly lower rate structure. The customer will appreciate the attention and may even reject future competitive attempts.

Success factors

A warehouse does not justify its investment simply by providing data that are already available or by confirming intuitive impressions. The warehouse needs to meet requirements that are more stringent, or it is not worth constructing.

Warehouses need to *identify the nonintuitive* explanation. If the warehouse only restates the obvious, it is not an economically viable tool. The value of the warehouse and its accompanying intelligence is that it seeks relationships that are meaningful yet overlooked. The software platform should be able to review more variables concurrently than any individual and sort through data on an iterative basis until it finds the answers buried in the database. The various techniques are engineered to find any relationship that bears closer attention. To the degree that the data warehouse finds no surprises, it has not met its obligations.

The warehouse needs *timeliness* to identify emerging trends, not historical fact. The market moves too quickly for backward analysis to be useful. In some cases, the warehouse will point to potential products and markets. It must do so in a manner that allows a telecommunications provider to plan, develop, and launch products and services before competitors can. The life cycle for all products is shortening, and it will certainly be shorter for telecommunications services in a fiercely competitive market. Providers need to offer products, grab market share, and pull in profitability before there are too many competitors fighting for the customer base and eroding share and profitability. The earlier that a trend or need is recognized, the more profitable the product.

The warehouse must prioritize the best opportunities when *isolating target markets*. The most significant relationships uncovered in the analysis are not necessarily the largest customer bases. Therefore, the analysis should rank the opportunities in a variety of ways to maximize the positive impact the results will have on future marketing.

The warehouse can help to provide advertising venues for niche services. Suppose there is a relationship between, for example, a consumer service offering and a seemingly unrelated hobby. This knowledge enables the provider to launch an advertising campaign in unusual venues (such as a hobbyist magazine) at an uncommonly low cost as compared to its impact.

Warehouses must demonstrate strong operational *performance*. The basic operational measures most frequently applied to applications that support operational business processes simply do not apply to warehouses. Data warehouses that measure only volume and efficiency benchmarks will not meet their business objectives. Instead, the performance measures can include level of integration, response time to users, and user surveys describing and quantifying value. At best, a warehouse audit conducted routinely could attempt to link marketing initiatives and results to the work performed in the warehouse. While a genuine cause-effect relationship is probably not verifiable, consensus can attribute some successes to information derived from the warehouse.

Pitfalls

While data warehouses offer significant benefits and will be competitively necessary in the long run, companies considering a warehouse system need to be aware of their risks. They also need to realize that there are unresolved problems related to the use of warehouses.

Warehouses combine the most difficult elements of information systems planning and development. Warehouses take data that were not developed with this kind of measurement in mind and combine them with other data they were never supposed to meet. Most often, warehouses are intended for use by individuals who are knowledgeable about neither the technology of the warehouse nor the intricacies of the statistical analysis. Warehouses by their nature do not have a specific or measurable data outcome. Telecommunications providers have historically collected an enormous amount of customer data and then used it only for billing, not for analysis. All of that data looks appealing in a warehouse plan, but whether it is useful or cost-effective in a warehouse is another issue.

Large information systems projects often grow out of scope and do not produce the intended benefits. This can take place even when the outcome is well-known, the users are technical, and the data are sound. Adding these uncertainties to a data warehouse project increases the need for significant management attention during system development and increases the risk of failure.

The cost of a warehouse is huge, and the value is difficult to measure. Data warehouses are not for the shy or the impoverished. The cost to implement a data warehouse is estimated at above $2 million [6]. This price will vary with the size of the database and the quality of the source data. Moreover, it depends upon the use of the data and the analyses performed.

The return on the investment should occur in two years, but this is a very broad measure. After all, most of the benefits are assumed to exist because of events that would not have occurred without it, a somewhat circular rationale. Most logicians would be mildly uncomfortable with that as a justification. Such benefits include new sales (how can one be sure that they would not have happened?), bills that were paid (only compared to what might have happened), and customers that did not leave (who did not necessarily try). Moreover, none of these agreeable events, when they do occur, happens in a vacuum. While the warehouse and the knowledge it generates are creating marketing opportunities, customers are also affected by the company's other marketing initiatives, the behavior of competitors, the customer's personal requirements, and the overall economy. Still, most companies building warehouses appear to be convinced that the impact on their businesses has been worthwhile. They are continuing to invest.

Building and maintaining a warehouse requires coordination and management investment of time. As much of the data are interdepartmental, the parochial interests of the participating managers can slow or derail data collection and maintenance requirements. Those individuals with access to the warehouse and a stake in the results need to invest time in learning about statistical functions—so that they know when results are significant—and to develop technical skills that they normally would not need. Many large information systems projects fail because training, documentation, and other implementation requirements are not covered, especially when budgets are strained. If it is not used, the warehouse is worthless. Warehouses are especially vulnerable because much of their usage is at the discretion of users and is not prompted by operational transactions.

Each customer's infrastructure might be difficult to compartmentalize. Pretend for a moment that you are building a warehouse. Your own company's data are perfectly sorted. You are certain that no records are duplicated or missing. You want to create clear profiles of each customer. You can get close, but you can never get all of it without the customer's active participation. For large business customers, many companies will maintain accounts by division or group. In part, this is a legacy of a regulatory environment and inactive marketing by telecommunications providers over time. The warehouse will not instantly identify opportunities in large account management unless the warehouse's owner finds the accounts first. This could be worthwhile, if volume discounts, service bundles, and large account management are among your skills. Such an analysis can uncover pockets of missed opportunity, where certain divisions are served by competitors. You can discover, for example, that, once the data have been scrubbed and sorted, your customer list includes regions 1, 2, and 4 of a Fortune 500 company. Region 3 is thus a probable and qualified prospect.

Data analysis can become another business excellence requirement. While it seems like an important differentiating factor, it is possible and even likely that the data warehouse could become necessary for a business to compete evenly rather than to surpass its competitors. The first company to offer employee benefits was undoubtedly deluged with the most talented labor. Now employee benefits are among many factors, and it is more difficult to excel cost effectively. Similarly, the tools offered by warehouses provide great benefits to the few who use them effectively. Once they are used rather universally, standards for the marketplace will rise, costs will reduce, and competition will increase. Those telecommunications providers that are not gaining the benefits that warehouses offer could be left behind.

SELF-ASSESSMENT—CUSTOMER PROFILING AND DATA MANAGEMENT

The following are some questions telecommunications marketers can use to assess the benefit of building a data warehouse.

- What business information will you require to compete successfully in your target markets?

- What is the quality of the business information you currently maintain?

- In how many organizational places do the data now reside? How difficult will it be to gather the data and consolidate them?

- How well can you compete with other providers in making the required investment if you choose to develop a warehouse? Are you prepared to make a large investment of capital, labor, and time?

- Who would the users of the data be? Are they capable of making the decisions that will provide the needed benefit and justify the warehouse's investment?

- Will a warehouse be necessary to compete in your market segment?

- Can you estimate the benefits to be gained by maintaining a warehouse or estimate the development costs?

- What data elements do you need to collect that you are not already collecting, whether you build a full-fledged warehouse or not?

References

[1]   Anonymous, "Telcos Getting Keen on Warehousing," *Telecommunications,* Americas Edition, Vol. 32, No. 8, p. 18.

[2]   Engebretson, Joan, "The Cost of Entry," *Telephony,* Vol. 233, No. 6, pp. 18–24.

[3]   Bushaus, Dawn, "Cold Storage Gets Hot," *tele.com,* Vol. 1, No. 4, pp. 58–61, 64.

[4]   Evagora, Andreas, "Profiles in Profiling," *tele.com,* Vol. 1, No. 4, p. 54.

[5]   Salak, John, "The Fix Is In," *tele.com,* Vol. 2, No. 7, pp. 40, 42.

[6]   French, Michael, "Mining for Dollars," *America's Network,* Vol. 102, No. 8, p. 24.

16

Customer Loyalty and Managing Churn

What is churn?

Churn refers to the annual turnover of the customer base. If 5% of existing customers leave, whether or not they are replaced, churn is 5%. The opposite of churn is customer retention. Obviously, a company's goal is to maximize customer retention and minimize churn.

One way to prevent churn is to create or maximize the costs of leaving. A switching cost is a cost that results from leaving a supplier, as perceived by the customer. One switching cost is the administrative charge that is applied by the customer's LEC when a customer changes long-distance providers. Carriers now routinely pay that cost on behalf of any customer who is willing to switch.

Relinquishing future bonuses can also deter customers from switching. Many carriers offer usage awards that accrue over time. A customer who leaves a service provider when earned awards have not yet been vested or redeemed will lose the remainder, which has economic value. Consequently, out-of-pocket charges and unredeemed benefits both act as switching costs to customers.

Customer loyalty can be viewed as the willingness of customers to remain with a telecommunications provider when a totally measured and rational decision would be to leave. In other words, the customer that stays with a carrier for five years has been retained but is not necessarily loyal. To be loyal, the customer needs to know that other providers offer better, cheaper, less risky, or in some way superior services. After all, a customer who stays with the best and least expensive supplier is not loyal, just sensible. The loyal customer wants to stay with a supplier because its competitors are exactly equal or for reasons that cannot be quantified.

The rate and cost of churn

Churn has been exceedingly high in those areas of telecommunications that are the most competitive: long-distance, wireless, and Internet access. Perhaps the level of churn for these services will subside when competition is more mature or when the services offered are more differentiated. Local service providers should expect significant churn for at least the first five or 10 years that all services are deregulated. While churn is often present in commodity industries where services are differentiated primarily on the basis of price, churn also occurs when customers are disappointed in the services they receive and believe that some other provider will be better.

Churn in the U.S. wireless industry has been approximately 30% annually [1]. While this rate is quite high compared to other industries, the annual 50% industry growth has kept the churn tolerable. Customers who leave are replaced immediately, and the revenues lost are not as evident. Still, the cost of that rate of churn is estimated at more than $4 billion a year in the mature markets of North America and Europe. For a perspective on how high that level of churn is, consider that many U.S. markets, over the study period, only had two wireless carriers and that many users were bound to a contract for at least a year.

In 1998, Strategy Analytics published a report stating that 38% of one wireless segment, the mobile family, was at risk to churn within a year, representing a potential loss of $4 billion in service revenue. This group (a household with at least one child 16 or younger) cited poor customer service as a reason for churn twice as often as did subscribers as a whole. Conducting an analysis like this one enables companies to focus on the attributes of their offerings that are most responsible for churn among certain groups. It is intuitive to assume that families churn because they are most sensitive to price because many consumers are price-sensitive. Nevertheless, these results suggest that an investment in the customer service quality itself is a better solution for these customers.

Companies sometimes create their own churn when they introduce disorder into their service and price structures. Andersen Consulting helped one European carrier discover that, of its 25% churn, only 5% of those leaving were defecting to competitors [2]. Another 10% were moving to different packages or networks offered by the same operator. While this type of churn is not as devastating as the complete loss of the customer, the effect on profitability is unfavorable because the loss and recapture of the customer incur virtually all of the out-of-pocket costs of normal churn.

More than half of all U.S. households have switched long-distance carriers at least once, motivated largely by price [3]. This suggests that long-distance service is still, to a large extent, viewed as a commodity by customers.

Some churn can be attributed to the transformation from a regulated monopoly to a competitive market. Some customers merely want to try a new provider, or they are lured by low promotional prices from a new competitor. Nonetheless, if customer curiosity about alternative providers were the driving force, churn would decrease as the market matured. In the residential long-distance market, according to Yankee Group research, churn rose more than 50% between 1994 and 1998 [4]. Other factors are apparently at work.

According to The Strategis Group, Internet access users create monthly churn at about five times the rate of paging and long-distance service, about 10%. Not all ISPs experience the same rate of churn, which can vary from about 8–30%. This compares with churn of 3% of pager users and less than 2% of long-distance telephone customers [5].

Some of the churn among Internet users was attributed to the wider array of choices available. The majority of the churn (about 80%) was due to factors within the ISP's control, such as busy signals, connection speed, and poor customer service. The same study determined that two-thirds of those who cancel service with one provider sign up with another access provider. This finding is important; these customers unquestionably represent revenues lost.

Nearly 40% of Internet users have switched ISPs at least once. Business users, at about 1–3% per month, churn much less than consumers [6], but those rates are still troublesome in this highly competitive market.

While Internet access is roughly comparable to other telecommunications-based services, there are several important differences to note and potential lessons to learn.

- The major Internet access providers offer free service for a limited time to new members. This causes some users to subscribe only for the no-charge period and then leave for another free offer or a less expensive provider. The lesson for telecommunications service providers is that free promotions generate customer sales, but the quality of the customer acquired does not equal that of those who are drawn by other means.

- Many Internet access services choices are available, and except for the top few providers, the services are near-commodities.

- While Internet service is a new business, customer expectations for quality and reliability are based on the last century of telephone service. This provides a lesson for local service providers, which will undoubtedly be expected to provide the equivalent quality of monopoly-provided services at lower and lower prices.

- Paradoxically, Internet service providers do have a high switching cost in the form of the user's e-mail address, which is bound to the domain name of the provider. Some entrepreneurs offer "lifetime" e-mail addresses, capable of forwarding to the current ISP used by the customer. Other Web "portals" such as Yahoo offer free e-mail accounts as a way to get users onto their sites. This mitigates the switching costs for those users who create the high churn. Local number portability is the equivalent capability (and is a regulatory requirement) in the local exchange service market.

The elusiveness of loyalty

Loyalty—to an employer, to a supplier, even to a family—has decreased throughout the century. While it is easy to assume that the decline of customer loyalty is a casualty of a decaying moral structure, there are many rational reasons for customers to be less loyal to suppliers [7]. The first reason is that customers have an abundance of choices. Instead of choosing based on loyalty among several products that each miss the mark, customers can now find the style and variation that meets their needs exactly. The likelihood that this ideal product is provided by their existing supplier is improbable.

Related to product diversity is channel diversity, illustrated by the emergence of superstores and new sales channels. Catalogs, the Internet, and huge retailers provide customers with access to a wide array of manufacturers and products. The corner store of 50 years ago undoubtedly carried only a few choices, at most, for any given product. Its chief offering was convenience, not variety. Customers were indeed more loyal, but their choices were limited. It can also be argued that the growth of suburban communities and the abundance of automobiles created mobility and indirectly affected the loyalty of customers to any particular institution.

Second, television, the Internet, and an assortment of print publications have made a tremendous amount of information available to customers. It is easier to learn about competitive choices, and even make comparisons on a feature-by-feature basis. Travel sites on the Internet already empower customers to optimize their travel based on price, schedules, or features of the hotel or airline. New sites are performing shopping services by locating the product the customer wants at the lowest price.

The third factor cited is entitlement, the customer's sense that choices are owed to them. This is undoubtedly due to the customer's ability to find an abundance of choices in virtually every area of consumption. The industry that provides no choice is an anomaly.

The fourth is commoditization, the blurring of differences between products. This refers to the manufacturing processes that create products with identical features, strict quality controls, and competitive response. When one company makes a popular product in plaid, its competitors follow.

The fifth is insecurity, the fear by buyers that loyalty is too expensive when one's own economic future is uncertain. The sixth cause is time scarcity, meaning that buyers have sacrificed product loyalty for the maximization of leisure time.

Another possible cause keeping customers from unqualified loyalty is a general perception that loyalty has eroded everywhere. Globalization and technology have created intense competition and enormous economies of scale. Companies cannot be as loyal to employees or even customers in the way that the old family-run businesses had been. Employees are more mobile for a variety of reasons, so employers cannot expect their loyalty either. Giant corporations, while providing many advantages to buyers, struggle to develop personal relationships with customers.

It is not impossible to gain the loyalty of customers, but it is by no means a guarantee. Loyalty is not inertia. An incumbent company's base of customers includes some subscribers who actively stay with the carrier in spite of competitive assaults, and others who stay passively. The customers in the latter group are vulnerable to any competitor that meets them more than halfway.

The benefit of these societal changes is that any of them can be used successfully by the skilled telecommunications marketer to gain a competitive advantage. Whether a superstore is a superior channel to a small shop for selling telecommunications services will be tested in the marketplace. Obviously, the largest providers have the capital to select any sales channels and offer a wide array of services. Niche providers will need to approach their markets with as much analysis as the giants to earn their customers' loyalty.

The importance of loyalty

Research sponsored by Xerox Corporation found that customers who reported that they are "totally satisfied" with a service are six times more likely to repurchase products than those who are merely "satisfied." Moreover, loyalty is profitable. The most expensive period in the customer life cycle is in their acquisition. The longer customers stay on, the less costly they are to maintain. Longtime customers bring in new business through word-of-mouth advertising, representing a distribution channel virtually free of charge. They are the most likely to buy additional products and services during the life cycle of the account. They are the most knowledgeable about business processes and the use of the product; they are least likely to create load on the support infrastructure.

It defies conventional wisdom, but it follows that the best use for limited marketing resources is to change the "satisfied" to the "totally satisfied," that is, the loyal, assuming that the needs of the "totally satisfied" have been adequately met.

Telecommunications providers have already learned of the value of investing to improve customer loyalty. The variables in Table 16.1 will help to quantify the value of a one-year customer versus a three-year customer.

The brief analysis presented in Table 16.1 makes broad assumptions based on loyalty research and is loosely based on research by Insight Research Corporation [8]. The model also leaves out several variables. For example, the one-year customer is more likely to leave during the first year than the three-year customer is during the third. Retaining a customer in the third year of service probably costs less than in the customer's first year. The provider will reasonably experience more uncollectibles from the first-year customer than in the later years of the

Table 16.1
Value of Customer Retention

	One-Year Customer	Three-Year Customer
Net profit—year 1	$500	$500
Net profit—year 2	0	500
Net profit—year 3	0	500
Total profit	$500	$1,500
Acquisition/retention cost	−300	−300
Customer value (net)	$200	$1,200

relationship. The analysis also omits the economic cost of money, but it still makes the point.

Wireless providers 360° Communications and Bell Atlantic Mobile say that customers who sign up with a provider-trained sales representative are much less likely to churn than those who subscribe at a general retail outlet [9]. The acquisition costs for these customers are lower, too.

The cost of churn and the value of loyalty can help provide the financial justification for building a data warehouse. (This topic is discussed in Chapter 15.) Furthermore, customer loyalty is an essential weapon to help prevent succumbing to price wars. Loyal customers do not switch providers for small or short-term price advantages.

Measuring customer satisfaction

One valuable way to learn to be the kind of supplier that attracts loyal customers is to measure customer satisfaction. There are many benefits, beyond simply earning more revenues, for making this investment.

- These measures help to define what the overall level of satisfaction is for your company and product and therefore deliver an assessment of the company's vulnerability.

- This research identifies either the individual or the profile of the customer who is not "totally satisfied," that is, the customer at risk.

- Customer satisfaction research explains the customers' point of view of the best and worst attributes of their experience with your company and products, not your estimate of their experience.

Many types of market research are available. Primary research requires the researcher to conduct new research to meet specific needs. Secondary research gathers a variety of primary research studies and draws conclusions or applies it to the situation at hand. Of course, primary research can be extremely customized, but it is much more expensive than secondary research. Primary or secondary research can be conducted in-house or contracted to outside specialists.

Customer research can utilize operations data for tracking customer activity, surveys, outbound phone calls, toll-free inbound customer service lines, service monitoring, post-transaction follow-up, interviews, and focus groups. Scripts can be added to daily customer transactions, as long as the customer is not inconvenienced in the course of doing business with the company. Customers can be encouraged to call to participate in surveys by offering free services, as long as the enticement does not distort the makeup of the respondent group. A telecommunications provider can offer a discount on the next bill for completing a satisfaction survey. Would only the most satisfied or least satisfied customers participate? The results will need to be normalized against a very representative population.

There is a science to survey design, and the survey developer needs to understand the precise goals of the market research. Customer perceptions are not equal to the trails of business processes. Survey results can mislead management when the survey misses the business issues.

Suppose a measurement tracked the repair of hardware that was returned for service. The customer called Saturday afternoon and sent the device on Monday morning for Tuesday delivery. The item was received on Tuesday and, on Wednesday, given to the repair shop, which fixed it in several hours. On Thursday, it was shipped back to the user for late Friday delivery. According to the repair shop, the item was turned over in less than 24 hours, but the user lived without this device for an entire week. Would the survey report excellent service? Can the shop's report be compared to a follow-up call to the customer? Will the customer expectation of 24-hour turnaround be met?

Mystery shopping programs have become a popular technique for understanding customer perceptions. Trained shoppers are sent to sales and service locations with the intention of buying or using the services and facilities of the sponsoring company. They produce reports that summarize their experiences. Mystery shoppers can evaluate sales or service personnel; measure queues, speed, or performance; and provide insight into the customer's perception of employee attitudes or the entire buying and service experience.

Financial and operational measures can assist to capture the value of a customer or segment. One valuable goal is to determine which customers are most profitable to keep. Companies can gain insight into customer behavior and customer profitability, but there are many areas to review.

Churn, of course, is a major influence on customer lifetime profitability. Other factors are important, too.

In telecommunications, network utilization data can be valuable in the aggregate to note trends and conduct planning. Utilization by individual subscriber can provide intelligence about the customer's potential for additional sales, or for churn. Grouping customers into segments and comparing their service usage to the average customer can provide insight into potential service bundles or promotions that will lead to incremental sales. An analysis that relates network investment and other costs to the lifetime revenue stream associated with those costs will help to target the most profitable customer groups.

Profitability analysis by customer type can help service providers avoid investing in unprofitable customers. Delinquency profiles can assist in recognizing delinquencies in their earliest stages and creating customer assistance programs to help customers meet their obligations in a structured manner.

Because existing customers are the best prospects, telecommunications providers benefit from making a detailed analysis of sales to the existing customer base. This analysis can be conducted with the assistance of the data warehouse.

Raising the bar

One of the realities of competition is that customers easily grow accustomed to the improvements brought about by competition. By the time one provider manages to raise quality or reduce costs enough to pass the benefits to customers, alert competitors alter their business plans to match the new playing field. Soon all remaining players are offering similar, yet improved, products, and within a very short period, customers have forgotten how the marketplace used to operate.

The emergence of quality programs has contributed to the rise in customer expectations. In the United States, superior-quality and cost-effective Asian products placed enormous pressure on American manufacturers to respond in kind. Quality programs became a competitive differentiating factor and business process reengineering a necessity to remove superfluous costs. One of the elements of many reengineering and quality programs was the notion of continuous improvement. Continuous improvement institutionalized the competitive nature of price and quality, delighting customers and adding urgency to providers' business life.

Retention techniques

Providers use a variety of retention techniques to maintain their customer base. Naturally, because the same customers are valuable to nearly all providers, these customers benefit by a spiraling competition of attention and awards. The amount of effort consumed for retaining customers is limited only by the determination and the resources of the provider.

Technological support can be valuable in improving customer retention. A new category of customer relationship management (CRM) software is available to help companies increase their sales and retain customers. CRM software supports a company's marketing strategy vis-à-vis an individual customer or account. Combined with other marketing systems, CRM technology enables companies to manage the complexity and abundance of information about a customer account and buying history. This can help to identify new sales opportunities or areas of vulnerability.

American Personal Communications has used intelligent call routing to improve service to customers and build relationships [10]. If a customer calls in for assistance on two consecutive days, the system routes the call to the same customer service representative. For all service calls, the customer account information appears on the screen for the representative, based on the likely reason for the call. Calls that come in within a few days of the billing date generate information about the bill, for example.

A loyalty program, or customer retention program, is one way that companies try to retain their customers and learn from the customers who leave. Loyalty programs contain several processes, including the following:

- Tailoring marketing and sales efforts toward potentially profitable and loyal new customers;

- Creating services and a customer care infrastructure to satisfy existing customers;

- Recapturing customers who appear to be vulnerable to competitive offerings;

- Learning about exiting customers' reasons for leaving.

Some customers leave because they no longer have a need for the product. Others leave for competitors. For those products and services that are purchased once, it is very difficult to know exactly if or when the customer has been lost. If a customer purchases a new telephone from a competitor, or replacement computer equipment, or new software, there is no record that the original purchase is no longer used. Computer and software companies use upgrade pricing in an attempt to retain customers. Of course, their competitors often match the upgrade prices with competitive upgrade pricing to eliminate the advantage of the incumbent supplier.

For subscription services, the provider is always notified in some way that the customer has been lost. In some cases, such as Internet access service, the customer needs to notify the ISP personally that the relationship has been terminated. In other cases, such as long-distance service in the United States, the customer simply contracts with a new provider and does not need to terminate the previous account. An exit interview with a departing customer can provide important information. If a subscriber does not need to notify the provider, the provider should seek the customer out for an exit interview.

It is essential to survey customers who leave to determine whether the cause of their departure was controllable. The survey can be simple and inexpensive, administered at the time the customer terminates the account, if possible. The scripted survey would first attempt to discern the reason for leaving, then entice the customer to stay, if possible. For providers that do not have an opportunity to consult the customer before the customer leaves, a postdeparture survey is still worthwhile.

Loyalty programs need to be targeted toward desirable customers. While this concept seems obvious, the ideal program should appeal to the profitable long-term customer and should dissuade others. Premiums after one year or five years are attractive to customers who do not expect to churn and can serve as both a reward and a switching cost. Promotional gifts or awards that do not compound over time will attract customers superficially, and a similarly superficial offering will draw them away in the near future, before the costs of acquisition have been recovered.

AT&T offers higher value awards to its high-volume residential customers and has added an administrative service charge to accounts that fall beneath a minimum calling volume. Most IXCs expect that bundled services help to encourage loyalty among the most desirable customers.

Loyalty program design should consider escalating the proportionate as well as the actual awards for the higher levels of purchase volume. A loyalty program that awards points of some kind can offer double points for usage over a certain number of minutes per month or double points after a subscriber has been loyal for more than a year. Pacific Bell experimented with an awards program that increased the award as the calling level increased.

Tie-in marketing refers to a promotion that involves separate products, such as a movie and a fast-food restaurant. The objective in tie-in marketing is that customers for one product can represent new customers for the other product and that net sales will increase. Generally, tie-in marketing is used for promotion, that is, short-term sales increases, but telecommunications providers have used tie-ins for customer retention as well. Some long-distance providers award airline miles for dollars spent on calling. Conversely, airlines offer free calling in their frequent traveler programs.

AT&T offered weekly pay-per-view movies in a tie-in with DirecTV, a direct broadcast satellite (DBS) provider in which AT&T had an ownership stake. The free movie presented a cost of switching to the customers, who would lose this bonus if they signed with any other long-distance provider. As this part of the offer simply adds value to AT&T's product, it strengthens the brand but does not necessarily promote long-term loyalty. However, in the same promotion, customers earned one month's free DBS service after a year of remaining with AT&T as the long-distance carrier, adding a long-term incentive to stay.

Providers can retain customers by creating switching costs. The first "frequent flyer"–like program was launched before there even was passenger airline service. Trading stamps were used by retailers since early in the twentieth century to encourage buyers to purchase more and to make purchases from one retailer rather than another. These purchase awards act as a switching cost for the customer.

Like any promotion, awards have a cost, and they can attract otherwise undesirable customers. Moreover, awards can become a competitive baseline that is absorbed into the cost of doing business. An airline or hotel without a frequent traveler program is at a significant disadvantage when it is otherwise equal to its competitors.

Similarly, providers that reimburse the customer in some way for lost awards will have more success than those that do not. Certainly, customers expect that administrative fees and even expenditures for required hardware would be covered by the company gaining the new business. Where reasonable, compensating customers for benefits lost can be classified as a customer acquisition expense. Whether this expense is justified depends on the profitability of the service and the expected life span of the customer relationship.

Whether the award should take the form of cash, bonus products, or unrelated merchandise is part of the design of the program. Merchandise awards have a value as trophies that cash (or cash credits) cannot match. The merchandise serves as a reminder of the customer relationship for a long time, in a way that a cash award, bonus products, or discount cannot accomplish. In addition, the customer is invariably working toward an award when opportunities to switch to competitors appear. The personal stake toward the award-in-progress can serve as a switching barrier for the customer.

Merchandise awards are not always the clear choice in loyalty programs, though. They require significant overhead to administer and are not as appropriate for or appreciated by low-income recipients.

Targeting preferred market segments

Several telecommunications providers, including AT&T, Sprint, and BellSouth, have invested in support organizations to help small business. While this initiative helps to engender customer loyalty, these centers also help small business owners find solutions that improve their business through increased telecommunications expenditures. Sprint, for example, provides a Business Solutions Center with 140 small business specialists.

Other providers target specific industries with service and awards intended to generate loyalty. As telecommunications providers become more sophisticated in their market segmentation, the targeted loyalty programs will undoubtedly become more specialized.

SELF-ASSESSMENT—CUSTOMER LOYALTY AND MANAGING CHURN

The following questions will help telecommunications marketers to determine whether their loyalty and churn management efforts are effective.

- What is the present level of churn? Is it rising, falling, or stable?

- Can you identify the customers who are most profitable or otherwise most desirable?

- Who are the most loyal customers? Why are they loyal?

- Which customers are least profitable or cause the highest churn? What can you do to minimize the damage to profitability?

- Which customers are more important to you than to your competitors? How far will you go to ensure that they are "totally satisfied"?

- What customer retention programs are in place? Do they target the proper customers? Do they work? Are they cost-effective?

- What processes are in place for continuous improvement to meet rising customer expectations?

References

[1]   Lockwood, Judith, "Study Predicts 'Epidemic' Churn," *Wireless Week*, Vol. 3, No. 34, p. 1.

[2]   Siber, Richard, "Combating the Churn Phenomenon," *Telecommunications, International Edition*, Vol. 31, No. 10, pp. 77–80.

[3]   Ernst, Daniel, "Consumer Services," *tele.com,* Vol. 1, No. 9, pp. 57, 58, 60.

[4]   Lawyer, Gail, "Rate Games Exact a Price," *tele.com*, Vol. 3, No. 3, pp. 19–20.

[5]   Spangler, Todd, "Service Improves, But Churn Is High," *Internet World*, Vol. 4, No. 6, p. 28.

[6] Bushaus, Dawn, "The Price That Isn't Right," *tele.com*, Vol. 2, No. 4, pp. 92–94, 96, 97.

[7] Schriver, Steve, "Customer Loyalty: Going, Going...," *American Demographics,* Vol. 19, No. 9, pp. 20–23.

[8] Levine, Shira, "Selling Smart," *America's Network,* Vol. 102, No. 17, pp. 20–24.

[9] McGinity, Meg, "Home Remedies for Wireless Churn," *tele.com,* Vol. 3, No. 3, p. 34.

[10] Bernier, Paula, "Wireless Support Systems Extend Reps' Reach," *Interactive Week,* August 12, 1996.

Part VI

A Marketing-Driven Company Infrastructure

17

Company Organization

The infrastructure

Telecommunications service providers do not exist solely as paper documents of incorporation and the services they provide. The elements of an organization include the work activities, business processes, organizational structures, values and policies, and incentives and the organization's supporting technologies.

The competitive telecommunications organization of the future will seek both effectiveness and efficiency. *Effectiveness* occurs when actions lead to desired results. *Efficiency* occurs when actions consume the least possible resources. Efficient and effective organizations eliminate the activities that do not lead to effectiveness and work to reduce the cost of those remaining and necessary.

Deregulation alters the priorities of work efficiency and effectiveness. In a monopoly, the organizational elements need to be categorically effective, and efficiency is desired but not at the same level of importance as their effectiveness. In a competitive environment, both efficiency and effectiveness are required, but neither is required at an absolute level. What is required absolutely is value, the customer belief that the service provided is worth its price.

Work activities are the tasks performed every day in the course of conducting business. In virtually all industries, these activities have changed substantially, primarily because of automation. Technology has transformed many blue-collar activities to computer-based tasks that are virtually indistinguishable from tasks performed by white-collar workers. The nature of most other telecommunications industry activities will change again due to deregulation. In highly competitive industries, the target of most work activities is the customer, and all actions are influenced by competitor actions. In contrast, monopoly activities were often directed at maintaining service quality and meeting the demands of regulatory authorities. More marketing activities will be necessary, as will product development activities. The activities directed toward regulatory compliance will diminish substantially. The role of service quality will be more complex but remain important.

Processes connect activities to each other. Most often, processes create information and transfer it for further processing. A worker creates a document, such as an order, then passes the order to other workers who perform the work to fulfill the order or other tasks. Processes can be manual or automated, efficient or inefficient. It is the never-ending task of competitive companies to seek out process inefficiencies and remove them. Telecommunications marketing processes include market development, customer acquisition, and customer care.

Work activities and processes operate within *organizational structures*, the reporting and working relationships among workers. Organizational structures can be efficient or inefficient. As white-collar tasks become more widespread throughout the organization, the opportunity for organizational inefficiency escalates. Organizational structures also define the relationship between one service provider's workers and the outside world, including customers, suppliers, agencies of all sorts, and the press.

Values and policies represent the written and unwritten rules about how employees are expected to act. This area of infrastructure will undergo one of the most important transformations of deregulation.

Values and policies include work environment attributes such as work hours, dress codes, the amount of structure in various jobs, the level of risk expected of employees, and internal social policies. These attributes vary from industry to industry, and they vary less within industries. There is no model for what values and policies will be in the future for telecommunications service providers in a competitive marketplace. Nevertheless, because the traditional providers are at one extreme of many attributes, it is safe to estimate that the values will moderate over time. Transforming the culture of the incumbent telecommunications provider has caused considerable challenges and some controversy.

Incentives represent the compensation structures and the system of rewards and punishments. Telecommunications providers that choose to develop a low-risk, high-quality culture will continue the present systems of raises, team bonuses, and systematic promotions. Telecommunications providers that prefer more risk in the corporate work force will offer high individual bonuses and will be willing to terminate workers who fail to meet their defined objectives.

Technology is a crucial element in the supporting infrastructure. Companies in all industries compete with each other's technology infrastructure; technology companies compete more fiercely. Computer hardware and software companies often boast the most sophisticated customer technical support infrastructures. Telecommunications service providers are already exploring technologies as a means to cut costs, improve customer service, and improve the workplace for their employees.

Transforming the cost profile

Marketers for telecommunications service providers have an enormous stake in reducing or reallocating their cost profiles, simply because their resource needs will grow substantially in a competitive environment. Telecommunications providers will all increase the marketing budget, but marketers will need to make difficult choices about where the resources are best placed.

Many telecommunications providers have thousands of employees, yet they do not know the costs of their business activities in a form that fosters strategic market management. In some areas, costing skills are superior. Historically, telecommunications providers have been meticulous in identifying various costing views for regulators and in computing the actual cost of certain customized services.

For the most part, though, in a monopoly, all costs are aggregated and divided among customers. It is necessary only to keep the total cost—not any individual element—manageable. In a competitive environment, each component is essential. Price pressures apply to all corporate costs, and product pressures apply to individual service lines. Without a cost-recovery guarantee, the only recovery mechanism is sales. Each cost needs to generate its own amount plus an adequate return.

For local telecommunications providers anticipating the influx of competitors, the cost structure is a source of concern. According to Forrester Research, local companies post revenues per employee at about half the rate of the long-distance carriers [1]. While the local loop is the most expensive portion of the network, it is not evident that the difference should be this wide. This statistic is complicated by the uneven relationship between local service costs and the prices they command.

Many companies have embarked on initiatives to identify their actual cost of doing business. One popular technique is *activity-based costing*. Activity-based costing breaks business processes, services, or any other chosen parameters into their most fundamental components. The technique measures each component's consumption and then computes the aggregate cost. Activity-based costing can demonstrate whether a service line is as profitable as it appears to be, or it can determine whether adding a single sentence to the bill would reduce customer service questions in excess of its investment. The process is detailed, but definitive, and requires management to know its objectives in advance of the data gathering effort. For companies seeking cost reduction, its results are often compelling.

Reengineering enjoyed a rush of popularity in the early 1990s. The controversial practice is still prevalent, although it is now part of the mainstream of management initiatives. One underlying concept of reengineering was that when a company's business changed significantly, its supporting business processes often did not, damaging its efficiency and effectiveness. Assumptions that made sense in an earlier time needed to be unearthed and revisited. Moreover, technology offered an opportunity to foster efficiency and effectiveness by enabling processes that were previously unattainable.

Reengineering utilizes organizational and technological solutions to accomplish its results. A reengineering initiative might accomplish the following:

- Automate paper transactions to improve accuracy;

- Eliminate low-value activities to reduce cost;

- Consolidate fragmented activities to produce higher quality;

- Identify sequential activities that can be conducted in parallel to achieve faster delivery;

- Decentralize activities to improve service to external and internal customers.

As an example, sales teams often complain that the last week before a proposal's deadline is too hectic and that proposal team members have inadequate access to sections of the proposal, other than their own, that they need to see. Applied to a typical bid process in a sales division, the results would include recommendations to implement groupware technology to ensure that the sales team always reviews the most current version of the proposal and has access to every section. Restructuring the activities in the bid process, providing team members with the most lead-time to contribute to the proposal, would improve the proposal's quality and reduce last-minute overtime. A reengineering study would identify tasks that should be consolidated and others that should be eliminated. Some reengineering initiatives have reduced costs dramatically and improved the quality of the business process products.

The controversy surrounding reengineering often arose from the misuse of the technique, as a weapon for eliminating staff without eliminating the work they produced. Reengineering was also blamed for the failure of many projects, most of which did not have sufficient management commitment or implementation discipline. Now that much of the fanfare has subsided, reengineering is sometimes credited for the increase in U.S. productivity in the mid-1990s.

Telecommunications providers have embarked on significant reengineering efforts in the face of impending competition. While many are quite successful, some penalties have been paid. US West scaled back its program when it was apparent that customer service was suffering [2]. BellSouth replaced a reengineering effort that was focused on cost reduction with a customer-focused strategy in which costs are secondary. After a 1995 downsizing, Ameritech was sued for its declining service by regulators in Wisconsin. Other companies have had more success with reengineering. GTE emerged from its difficult reengineering program with a more streamlined customer service infrastructure and a network-monitoring product that it can sell to private companies. NYNEX's reengineering program provided installers with handheld computers, which let them activate service features and answer customer billing questions, eliminating work for other divisions and improving customer service.

Efficient organizational structures

As in so many other areas of management, technology has enabled companies to analyze the efficiency of their organizational structures. Metrics such as span of control, levels of management, and cost-to-manage demonstrate clearly the opportunities to reduce an organization's cost without impairing its effectiveness. *Span of control* refers to the number of subordinates under the management of a supervisor, or the average number of subordinates to all supervisors in an organization. Generally, field organizations can sustain 10–12 subordinates per supervisor. Organizations needing additional supervision will support smaller spans of control. A span of control that is too large provides insufficient supervision to workers. Too small an average span of control results in too many supervisors and wasted salary expenditure.

The metric regarding *levels of management* refers to the number of individuals between the top manager and the least senior worker. After divestiture, both AT&T and its divested offspring eliminated several levels of management to prepare for the competitive market. Telecommunications giants are not the only providers threatened with unnecessary costs in too many levels of management. Startups sometimes reward employees with supervisory promotions and supporting workers. If the mid-sized organizations reach a plateau and do not achieve the same level of growth year after year, some supervisors, and potentially whole levels of supervision, are unnecessary. This can be gravely dangerous for smaller companies. The telecommunications market leader with 10 levels of management instead of a more efficient nine will have fewer long-term consequences than the startup with three levels that should have two.

Cost-to-manage is a ratio that calculates the salaries consumed in the management of workers in an organization. The cost-to-manage is affected by span of control because a manager's salary divided by 12 is a much lower number than the same salary divided by four. Levels of management affect cost-to-manage as well because the managers of supervisors are part of the management cost. The more levels of management, the more salaries (and generally higher salaries) are added to the cost. Cost-to-manage is also influenced by the width of the gap between worker and manager salaries. Last, cost-to-manage is affected by the overall level of salaries paid to supervisory management. Higher salaries present a higher cost-to-manage.

One way that companies have addressed the need for workers to do more with less is through *empowerment*. Workers are provided the tools and the authority to make decisions at the lowest levels of the organization. Technology provides the workers with the information they need to make sound judgments, and the need for supervision is minimized. In a customer-focused organization, empowering employees to resolve customer concerns independently and on the spot improves the customer's experience. Customers prefer to work with a single employee in a single transaction. Empowerment makes that possible for more transactions.

Effective organizational structures

Like most companies, monopoly telecommunications providers traditionally organized to the needs of their businesses. While that was appropriate for its time, those needs have changed, without commensurate changes to the overall organizational structures. Telecommunications monopolies organized first by business function, such as construction, accounting (later information systems), customer service, and regulatory functions. When wireless technologies arose, regulatory requirements mandated separating the new services from the core business entity. Wireless and then other unregulated ventures were assigned to separate subsidiaries.

Along the way, recognition of the importance of market segmentation created new divisions to serve targeted markets. This signifies the beginning of a transition to a customer-focused organization. The business units are generally limited to marketing functions and often do not include service provision or supporting infrastructure. In their efforts to focus on serving specific markets, the marketing functions will compete for corporate resources held by the other organizational divisions. Slowly, the major telecommunications providers are moving to structures that support numerous service lines beyond wireline voice communications, where customers rather than regulators are the primary focus, and technology builds bridges when business processes are aligned.

Unfortunately, many legacy organizations, including RBOCs and AT&T, support vestiges of structure to meet regulatory needs, not the customers'. While AT&T and its divested subsidiaries are clearly moving toward a market-based organizational structure, all of these companies, three years after the Telecommunications Act was signed, maintain separate organizations for different service lines. The effect of vertical organizational structures is that the telecommunications providers' own infrastructures will prevent them from the customer focus that they will need to stay competitive. Customers want bundled services, but existing wireline and wireless billing systems are separate. Many customers want a single point of contact for their telecommunications services, but these vertical structures maintain separate service delivery, sales, and customer service organizations. Customer data collection is fragmented over a variety of databases, potentially without the ability to be reconciled against each other. When a company sells a bundled service or claims to be a one-stop shop, customers will expect the back office operations to be coordinated and invisible to them. In the present organizational model, this will be extremely difficult. Routine decisions affecting several service lines in a vertical model have to go to senior management for resolution, potentially introducing delays and political wrangling that should not occur in a simple customer transaction.

One barrier to the expeditious move to a new organizational model is that the transition will be disruptive and costly. Two or more information systems will need to be consolidated, sales forces will require retraining, and executives will be reshuffled. Removing organizational barriers and duplication is nearly as monumental as a merger would be, and just as necessary. Another barrier for some companies is the continuation of regulatory restrictions on their businesses.

Company culture

The introduction of competition in long-distance, followed by the 1984 AT&T divestiture and industry deregulation, imparted a dramatic jolt to the culture of the monopoly environment of U.S. telecommunications. Monopolies operate under different pressures than do competitors. Monopolists are internally focused and internally competitive. Competitors look to customers and competitors to provide insight into their performance.

Monopoly management tends to be risk-averse. Employees and managers in monopolies usually forego the career opportunities offered in small growing companies and avoid the cutthroat but lucrative cultures of larger enterprises. Monopolies support a larger employee base than their competitive counterparts, but the demands on quality are very high. Employees of monopolies, like those of government, are highly loyal. Turnover is minimal. Monopolies can afford to rotate employees throughout the company in the understanding that the on-the-job training will most likely be amortized over a long career.

In short, employees of monopolists have different needs, different work patterns, and different approaches to risk than their counterparts in competitive enterprises. How does the monopoly move the culture to the competitive arena without losing the values and culture that created the quality and reputation of the existing company?

Bell Atlantic recognized that a cultural gap existed, and that employees required a reorientation to a competitive world. It created the Bell Atlantic Way program to revise the conventions of employee behavior. The company trained 65,000 people in 2.5 million training hours. Like other companies that empower employees to make rational business decisions, Bell Atlantic viewed its people as associates. Associates were encouraged to "coach" each other using nonthreatening language and symbolic rewards, such as "blue chips." The use of new terminology acted as a focal point to shift the culture through behavior as well as attitude. Bell Atlantic adopted a concept of "best cost" to ensure that the cost profile was improved but that quality and the company's reputation consequently did not suffer.

Knowledge transfer

The rise in information-based white-collar workers and the rapidly changing environment have created the need for knowledge transfer. *Knowledge transfer*, sometimes referred to as knowledge management or knowledge creation, comprises the collection, interpretation, management, and communication of knowledge. A robust knowledge in-frastructure offers the ability to retrieve knowledge anywhere in the organization whenever it is needed. Knowledge management programs can offer training and support to inexperienced staff and provide standards for decisions in similar situations. Managing knowledge through infrastructure minimizes the organization's dependence on a single expert or several experts in disagreement about solutions. Knowledge systems can act as trainers, help desks, decision support mechanisms, or diagnostic systems.

Because of its focus on storing and sharing information, technological infrastructure is an important aspect of knowledge transfer. Other fundamental organizational changes are required as well. US West established a knowledge program in its Network Service Assurance organization and its Local Network Operations that resulted in major efficiency gains [3]. The company has organized meetings between workers in the best performing organizations and those from underperforming teams. AT&T has created knowledge systems to capture knowledge and share it with other groups within the company.

In a competitive environment, knowledge transfer will increase in importance. Turnover of the most knowledgeable and valuable workers will inevitably increase as companies become more competitive, and as newcomers pursue senior employees. The expansion of geographical territories will make it impossible to have the physical presence of a particular knowledge worker wherever one is needed. As telecommunications services become more sophisticated, customers will have problems, and knowledge management systems will prevent the same problem from being solved from ground zero repeatedly. The rise in telecommuting will make it essential to find support for workers who cannot meet in person with the knowledge worker they need to reach.

Merger and acquisition integration

Consolidation is rampant in the telecommunications industry and shows no sign of abating. *Critical mass* is the term generally used to describe the size a telecommunications provider needs to be for success in the market. At the news of a merger of telecommunications giants, industry analysts who continuously mention the objective of critical mass have yet to assert that any provider has thus far achieved it. A deregulated telecommunications marketplace in the information age is unprecedented. Nobody knows the ideal size or scope for a telecommunications provider. Until the market forces stabilize, market participants will continue to merge and grow, if only for competitive parity.

Acquisitions are popular with both management and shareholders, but most of them fail. Sensing the convergence of telecommunications and computing, AT&T acquired NCR and divested it several years later. For the same reason, IBM made several significant investments in telecommunications companies and later sold them.

Management views acquisition as an entry strategy into new markets with less risk than startup efforts. Companies merge with their counterparts to expand territories as did Southwestern Bell Communications and Pacific Telesis or Bell Atlantic and NYNEX. Companies integrate horizontally through acquisition, demonstrated by AT&T's purchases of cable provider TCI and local business carrier TCG. WorldCom purchased MCI to expand its name recognition, customer base, and marketing skills. Some of these efforts will succeed, some will fail, and some will be immeasurable. The resulting marketing organization can increase the chances of a merger's success or be its downfall.

During the due diligence process (when senior managers on both sides are evaluating the merger partner), marketing management should participate on the team. Most organizational divisions have little access to the information they need to make important strategic, operational, and financial decisions. Due diligence is a learning process. Merger partners are reluctant and often forbidden to share information that would help the partner make decisions about the potential merger. One reason for the lack of communication is that if the merger does not complete, a strategic competitor would have valuable competitive information to use against its former partner. A second concern is that some laws prevent companies from sharing certain information and plans in private.

Marketing is a most important function in the information-gathering process. Competitive intelligence efforts are most often led by marketing professionals. Before merger negotiations begin, an astute telecommunications provider already knows the market strengths and weaknesses of the acquisition candidate and the potential positioning of a combined company. Competitive intelligence professionals can help to identify strategic partners and screen candidates identified by others. Marketing information is usually the easiest information to get about a company, as a telecommunications provider's objective is to impart its marketing message. More to the point, one partner's marketing finesse is often the reason for the other partner's interest. Involving the marketing function in the merger process will increase the likelihood of a successful union.

After the merger, the difficult and necessary process of merging corporate systems includes the marketing infrastructure. Customers will require company systems and processes to be seamless, and marketing efforts are the most visible to them. Inertia, culture, and scarce resources will discourage their prompt integration, but the sooner the company immerses itself in its new marketing identity, the higher its chances to beat the odds of merger failure. Alltel Communications acquired wireless operator 360° Communications and immediately evaluated the systems both companies were using. In the billing and customer care area, one partner has a long-term contract [4]. The system conversion will occur at the termination of the contract. While this situation provided significant financial consequences of merging operations, most decisions of this sort require less investment and more fortitude.

Benchmarking

Benchmarking has become a popular management initiative for improving the organization's effectiveness. The technique of benchmarking compares the performance of two or more environments while controlling the most significant variables. The goal is to establish which variables significantly increase the organization's performance. Variables can include the use of technology; staffing variables such as number of staff, skills, or training; work tools; procedures; and compensation plans. When two comparable organizations exhibit significant differences in performance, management evaluates the variables to determine if they are responsible for making the difference.

Assigning a causal relationship through benchmarking is worthwhile when it is accurate, but it is easy to draw conclusions based on the wrong evidence. A solid benchmarking effort requires planning, vigilance, and normalization of the data gathered. Like most worthwhile improvement programs, benchmarking requires commitment and investment.

Telecommunications providers can benchmark internally or partner with other providers to develop industry benchmarks. Marketing departments with a new service to sell can select several comparable test markets and use a variety of selling techniques. Single or combination approaches of direct marketing, print advertising, broadcast advertising, or promotions can be assigned to each test market. With a sophisticated CRM system, benchmarking can be used in a single market. At the end of the test, if enough data are collected to ensure that the groups are indeed comparable, the analysis can determine the effectiveness of each marketing approach. Later, the same customer data can be tested for churn or the customer's purchase of additional services. Using statistical techniques, benchmarking can demonstrate which selling approach is favorable in demographic groups of the same makeup as the test group. As telecommunications providers take their local service offerings nationwide, they will need to choose among a variety of marketing strategies, all with different costs. Benchmarking in test areas will confirm which approaches are most cost-effective.

Companies also partner with each other to develop benchmarks in critical business functions. A worldwide study of service development benchmarks in 1996 included more than 30 wireline and wireless service providers [5]. The effort demonstrated significant differences among participants; best-in-class companies significantly outperformed average companies in all aspects of service development.

Learning one's own position as compared to the best performers is only half the battle. An associated study of *best practices* isolates the behavior of the highest performers responsible for the improved results. A telecommunications provider that learns that its service development cycle is average must first decide whether its marketing strategy requires it to be better than average. If the service provider has decided to be a service leader, then the cycle time must be reduced. A look at the business processes of the cycle time leaders will provide ideas on the changes that will improve performance.

While industry consortia such as the one sponsoring the 1996 study are valuable, they are not always available. Market research firms can conduct similar analyses using primary or secondary data. Ideally, faced with a worthwhile benchmarking target, the market researcher will be willing to host a study on behalf of several providers. Sharing the cost of customized research with other telecommunications providers can be efficient and constructive.

Benchmarking can involve dozens of participants or only two partners. In fact, the benchmark partner does not need to actively participate or even know of a service provider's benchmarking efforts. An effective competitive intelligence program can provide enough information to estimate marketing benchmarks such as cost of a sale or churn. Using the competitive approximation as a benchmark, the service provider can review its own processes and effectiveness against the estimate. Similarly, companies that are not competitors but share business functions often learn creative solutions through benchmarking partnerships. A telecommunications provider eager to learn best practices for developing a compensation plan can partner with a world-class marketer of nontelecommunications services in exchange for sharing one's own skills in network management or scheduling repair calls. Service providers can benchmark their internal telemarketing operations against those of their service bureaus.

SELF-ASSESSMENT—COMPANY ORGANIZATION

Telecommunications marketers can use the following questions to begin an assessment of their organization's readiness for a competitive environment.

- What is the present state of your company's organizational infrastructure? Do gaps exist between current and ideal costs, culture, quality, or effectiveness?

- What is the present state of the infrastructure of the targeted competitors? How quickly do any gaps need to be closed?

- What are the tradeoffs involved in a cultural transformation? What changes are mandated? How far can the culture move to gain competitive advantage without losing other competitive advantages?

- Is there a process in place to spot and correct downturns in your company's cost profile, the robustness of its technology, or its quality before they are visible to customers or shareholders?

- Does your company support a knowledge management system? Are employees impaired in their effectiveness because needed information is not available when they need it?

- What impact would mergers and acquisitions have on your company's infrastructure? Is there a process in place to evaluate this impact before these important decisions are made?

References

[1] Wilde, Candee, "Growing, Growing, Gone," *tele.com*, Vol. 1, No. 2, pp. 44–47.

[2] Wilde, Candee, "Shrink Raps," *tele.com*, Vol. 1, No. 2, pp. 40–43.

[3] Docters, Robert G., Dan E. Aks, Reggie Van Lee, and Stephen R. Mintz, "Putting Knowledge to Work: From Abstract to Awesome!," *Telephony*, Vol. 231, No. 20.

[4] Meyers, Jason, "Confounding Connections," *Telephony*, Vol. 235, No. 18, pp. 20–24.

[5] Ogawa, Dennis, and Laura Ketner, "Benchmarking Product Development," *Telephony*, Vol. 232, No. 4, pp. 34–38.

18

Strategic Technology

Technology as a marketing weapon

MCI created the most recounted marketing weapon in telecommunications marketing history in 1990 with its consumer-targeted Friends and Family program. At the time the program was introduced, MCI was a very small competitor to market leader AT&T. AT&T had nearly every advantage: name recognition, perceived service quality, and customer inertia. MCI could not endure a competition based on its small discount on price. The Friends and Family program was created to increase market share.

The heart of the program was to recruit customers to sign up additional customers. MCI allowed each customer to identify a list of friends or family, also MCI customers. Any time the customer called a member of their list, a generous discount was applied to the price of the call. This discount encouraged consumers to recruit new customers for the discounted calling. MCI seized significant market share from its rival AT&T in a tactic to which AT&T could not possibly respond.

Part of MCI's genius was to attack AT&T on its own strengths and on its weakness. AT&T could not introduce a similar discounted program; virtually every consumer was already an AT&T customer. A Friends and Family program offered to AT&T customers would not gain enough new customers to offset the substantial decrease in revenues that the discount would cause.

In a practical sense, AT&T could not respond anyway. The introduction of personal lists complicated the call-measurement activity for billing. MCI, with new billing platforms and an emphasis on strategic information systems, could modify its systems with little difficulty. AT&T's legacy billing systems would have required massive modification at a huge cost. MCI's inclusion of the calling circle list was strategically important, although it added to the complexity of the data systems modifications. A customer calling an MCI subscriber who was not on the list did not receive the discount. Thus, as MCI's market share grew, the calls eligible for the discount did not grow as quickly.

The Friends and Family program, now imitated by carriers around the world, also provides MCI with some protection against churn. Participation in a calling circle constitutes a switching cost for the customer considering churning as well as for all the other members of the list.

As MCI's share grew, its Friends and Family program evolved into a more traditional discount program. Its focus shifted appropriately from customer acquisition to one of customer retention.

There are two lessons for telecommunications providers in the Friends and Family story. The success of Friends and Family demonstrates how information systems, invisible to customers, can support products that appeal to customers and cannot be matched by competitors. The second, more troubling lesson is that no provider, including MCI, has outdone that use of technology in nearly a decade in this technology industry.

Impact of IP telephony on marketing

The marketing of telecommunications services is affected by the technologies underlying the product. When wireless networks were upgraded to digital technology, it improved the service quality, enabled providers to add digital information services, and provided an enlarged arsenal for marketers to differentiate their services. Today's wireline telecommunications network is primarily a circuit-switched architecture designed to carry voice-grade communications. IP telephony is challenging this architecture in both capability and cost. Initial IP offerings required customers to have personal computers equipped with modems and install software that provided calling only to other holders of the software. Both of these barriers have been overcome by newer technology. Traditional telecommunications providers are monitoring the market conscientiously, and some have begun the process of migrating portions of their networks to IP technologies. Competitive local service providers such as CLECs and ISPs have embraced the new architecture as a lower cost, independent entry strategy to reach cost-conscious consumers and businesses. In 1998, about half of ISPs and CLECs anticipate offering voice over IP services in the near future [1]. AT&T and Sprint are offering IP telephony on a limited basis [2].

While the most significant impact of IP technology will be on the delivery of telecommunications services, IP telephony will also play a large role in telecommunications marketing. Long-disdained for unacceptable voice quality, IP services have improved considerably. They are expected to be equivalent in quality to the public switched network in the near future. The pricing of IP services, particularly long-distance services, is unencumbered by access charges or subsidies. Whether this is an unsustainable price structure will depend on regulatory decisions over time.

IP-based telephony services have already shaken the foundations of telecommunications marketing. First, the low per-minute pricing has placed additional pressure on traditional long-distance providers, if not to lower their rates further, then at least not to raise them. Second, the digital nature of Internet-based telephony makes its bandwidth usage more efficient than traditional voice networks and has the capability to include enhanced services beyond those already offered on the public switched network, again at a very low cost. These services include e-mail messaging, fax messaging, universal in-boxes, and fax broadcasting. Service enhancements like these will apply quality and pricing pressure to the marketing efforts of the wireline carriers. Third, IP telephony eliminates many of the distinctions that have enabled price discrimination as a foundation of traditional telecommunication services. IP makes no distinction between voice, data, or video traffic. IP does not differentiate between local, long-distance, or international calling. IP is not wedded to local numbers or routing arrangements. The only significant cost element is a user's access to the network. This can simplify pricing, and it can dramatically shake the assumptions underlying telecommunications rate structures. Fourth, the IP telephony providers are demonstrating an eagerness to partner with the traditional carriers, although regulators have rebuffed these attempts when they involve RBOCs. Last, IP telephony will bring "chat" to Web sites, and eventually video, introducing another very personal form of customer service. Chat enables an Internet user to call on a customer service representative in real time to assist the user in navigating the Web site, reviewing a personal account, or ordering service.

Furthermore, these untraditional providers are experimenting with pricing plans that were unthinkable five years ago. USA Talks.com announced a telephone-to-telephone Internet-based long-distance service for $60 a month. Whether the specific telecommunications provider will be able to sustain that price level, the landscape will be changed forever. Sprint's $25 per month flat rate weekend offer will not have the same impact, even though it is offered by a better known company. Sprint's weekend-only offer will not jeopardize the carrier's profitable business revenue, and at worst, it will reposition extraneous traffic from its busiest periods to times of overcapacity. USA Talks.com's Internet telephony service leverages Internet technology and has the opportunity to create a market for unlimited flat-rate calling if it survives the clutter of the marketplace and the cost of creating a market presence. While local providers are hoping for more measured service in their local markets, the marketplace could be creating contradictory expectations. In the end, the marketplace prevails.

Back-end systems

Northern Business Information, a market research division of McGraw-Hill, conducted a study comparing investments in operations support systems [3]. This study predicted that by 2001, customer care and billing would constitute more investment in the United States than in the operations, service provisioning, and planning and engineering support systems. Worldwide, telecommunications service providers will increase from their 1997 expenditure of $7 billion to $16 billion by 2002 [4].

The majority of telecommunications providers need to do something dramatic. Most of the largest incumbent carriers support legacy information systems that constitute a burden, not a strategic asset. AT&T has separate systems for each product line, including a system for billing private-line services and another for toll-free services [5]. According to *Information Week,* of the 21 industries surveyed, the telecommunications industry is in the top three spenders on information systems but is last in spending on new technologies. Other disappointing rankings include the next-to-last spot in research and development, third-from-last in Web sales and support, and fourth-from-last in electronic commerce sales. Legacy systems probably account for the industry's high ranking in both overall expenditure and year 2000 conversion, which is essentially a software maintenance function.

SNET has integrated many of its customer-related information systems to support the bundled services it offers. Productivity improvements have been estimated at 10–20% in the first year. US West has taken a similar approach to managing bundled services. While the transition to a single billing platform is taking place, the company presents a seamless appearance across multiple software platforms.

Satellite mobile carrier Iridium faced a unique customer care challenge because its product is by nature available anywhere in the world. Iridium supports several virtual call centers, including access number-directed language support in 13 languages and access to 40 others [6].

Most of the carriers' investment will center on marketing-critical systems such as billing, customer care, and data warehouses, covered in Chapters 14 and 15. Other technologies are also used to increase coverage, reduce cost, and improve customer service.

OSSs include several crucial business functions and represent an opportunity for telecommunications providers to differentiate their services. Besides the billing function, OSS applications cover the pre-ordering and ordering process, provisioning, maintenance, and repair functions. For telecommunications providers eager to serve the wholesale market, OSS capabilities are part of the product line. For RBOCs, prohibited from the long-distance market until certain conditions are met, OSS excellence sits squarely on the route to competition. For telecommunications providers whose strategies require superior customer care, OSS is fundamental to the goal.

SBC converted manual order-taking into a computerized sales process when it integrated several operational applications of its own and its merger partner Pacific Telesis [7]. Customer service representatives can access data across applications and across geographical boundaries. The system provides CLECs with the same access to network information as its internal system users.

GTE has established a program for customers to enter access service orders electronically through dial-up access and eventually a Web-based platform [8]. Similar provisioning platforms are in use at Interpath (local provider owned by utility Carolina Power & Light) and NBTel (New Brunswick, Canada).

Cable modem service provider MediaOne automated subscriber sign-on, service selection, and provisioning for its 100,000 Internet users [9]. Its goals were to lower costs, provide flexibility for market trial data collection, and improve the customer's experience. Eventually, the company hopes to reduce costs further by creating a self-provisioning environment in which the user can purchase the modem at a retail outlet and engage in the registration and provisioning activities with minimal assistance from the provider.

Electronic bonding refers to the electronic interchange of information between telecommunications providers and other carriers or their customers. The concept is not revolutionary; companies, including telecommunications providers, have supported electronic data interchange (EDI) for decades. Barriers to full implementation include the preponderance of wireline provider legacy software, a need for industry standards for data transfer, and a cultural practice not to share network information with outsiders. Telecommunications providers are also concerned that security needs to be sufficiently robust that customers cannot view internal proprietary network data or the data about other customers.

Interactive voice response

Interactive voice response (IVR) enables users to interact with a telecommunications service provider without involving a service representative and without requiring any equipment other than a telephone. IVR technology supports customer service applications by enabling users to conduct routine business without requiring an agent. Users can change administrative options; request information to be delivered by voice, by mail, or by a return fax; or provide self-service diagnostics and solutions.

While some customers prefer IVR's convenience to personal service, others resent having to use the technology. Poor menu designs in some systems or complicated applications can cause a customer to endure a difficult transaction. The best designs immediately instruct the caller how to reach a live operator.

Surprisingly, telecommunications providers do not use IVR technology to the same degree as their own customers in other industries do. While IVR is less sophisticated than computer-based alternatives, modem-equipped computers have not penetrated the user base. All customers have access to telephones.

All prepaid telecommunications platforms use IVR technology to identify customers and route their calls. Some allow customers to refresh the value on the prepaid card using IVR transactions. Consumer products companies have hosted promotional sweepstakes using IVR to build brand awareness. Financial institutions use IVR to link customers with service representatives to obtain emergency cash [10]. Airlines and airports use IVR to provide flight information.

Some telecommunications providers have applied the technology to marketing and customer service. Corporate customers of SBC can enter their account numbers through an IVR system. They are routed to call centers equipped with specific business applications.

Sprint introduced an IVR system that enables customers to check on payments, confirm account balances, and notify the carrier of special payment needs. Systems are available that enable customers to use voice response systems in which their responses to a series of questions can report trouble or determine the status of a problem.

Sometimes telecommunications providers utilize voice response in the product itself, such as prepaid phone card administration. Voice response systems are used extensively to reduce the cost of directory assistance services by postponing operator intervention as long as possible. IVR technology is also used in low-cost services such as automated collect and dial-around long-distance to eliminate the operator intervention, reduce cost, and make the service more attractive. Ameritech created a privacy product that utilizes voice response technology. The Privacy Manager service acquires the caller's name from Caller ID records or uses voice technology to capture and repeat the identification to the customer. The customer can then choose to accept or reject the call. This could be ironic if Ameritech takes the telemarketing approach of its long-distance competitors when the market is fully open for local service competition and discovers its prospects block its marketing efforts through its own resold technology.

The call center

Call centers have been a fundamental part of telecommunications marketing for decades, but the convergence between computing and telecommunications has made call centers even more valuable. Computer-telephony integration (CTI) is enabling call centers to be more efficient, more effective, and more responsive to the needs of both customers and service representatives. Automated call distribution (ACD), once used primarily to distribute calling volume among agents, now enables agents to work from remote locations. Information gathered before the agent is connected includes call abandon rates and calling volumes by time-of-day.

Early technology created the screen pop, a screen containing customer information that could be generated either through caller ID or through input of the customer account number. Newer systems route the caller to the right service representative based on language or geography, product expertise, or follow-through of an open case.

Long-distance carriers use outbound call center marketing so extensively in the United States that it has nearly become folklore. For low-volume consumer service, telemarketing is the proper sales channel.

Telecommunications providers also use inbound marketing for sales and customer care.

When customers call AT&T for customer service, the customer's number is matched against a database. The customer representative can see information about customers immediately, such as what language they speak and their bill payment history.

Case-based reasoning has been used by US West, AT&T, and BT [11]. Scripts direct the agent through a diagnosis of the customer's problem. The system leads the agent to documentation containing probable solutions. This technique will serve as a precursor to Web-based customer service, potentially eliminating the agent in most service transactions.

Newer technologies will link call center results to a company's intranet. Customer data will enable marketers to analyze the results of promotions as they are happening and make necessary adjustments. Marketers will see the service impact of new products by viewing customer complaints or questions within minutes of their occurrence.

Marketing support using intranets

Telecommunications services providers have found that intranets offer a cost-effective tool to reach employees and others who require special access to company data. MCI's intranet contains company data, links to the home pages of its competitors, news, market research, and links to valuable Internet resources. Its Business Markets sales and marketing staff can train themselves online [12]. These classes now account for 40% of the group's total training and have yielded more than $100 million in total company savings in four years. Using the intranet to answer customers' questions saves $12 million annually [13] and improves the responsiveness and quality of the customer service transaction. More than 10,000 Bell Atlantic customer service representatives refer to the company's intranet for answers to customer questions.

Telecommunications service providers are testing the ability to send alphanumeric pages through their intranets, offering wireless access to field service personnel and enabling customers to review their current calling plans in comparison with their in-house alternatives.

Marketing on the Internet

Telecommunications carriers are beginning to use the Internet as a sales channel. The first application embraced by most telecommunications providers was billing. Many carriers now have some billing presence on the Web, which provides competitive parity, and results in cost savings, up to $3 per bill [14]. Some of the savings are often passed to consumers in the form of pricing discounts. Most telecommunications service providers offer directory services on the Web.

MCI's Web site enables customers to view their accounts, create an online address book, send a page, or schedule a conference call. AT&T allows customers to enroll in its IP telephony service, view account information, send a page, create a computer-based conference call, or manage an AT&T Universal Card account. Sprint's site lets consumers review their account, order a calling card, or change their calling plan. US West customers can order service, move or add features, or disconnect their local service online, as well as post repair orders or receive technical support for Internet access. Users can type in their telephone numbers and learn online if their central office is equipped to provide the services they want. BellSouth customers can view their bills, order service or telecommunications equipment, purchase computer applications training, or buy company-branded gifts. Frame relay operators AT&T, CompuServe, MCI, and Sprint offer Web-accessible reports that present network management data [15].

Innovative use of the Web also originates with emerging carriers. Wireless provider Teligent launched a Web billing service to let customers pay bills online and sort calls by categories such as accounting code, destination, employee, and originating number [16]. This feature will provide a competitive advantage in Teligent's target market of small- and medium-sized businesses, which rarely support such capabilities internally but are more cost-conscious than many other segments.

MediaOne uses e-mail to answer questions about its cable and Internet services [17]. Artificial intelligence improves the efficiency of the system by parsing the requests and providing answers to the most frequently asked questions. This feature reduces the time that customer service representatives need to spend on most questions. SBC Internet

Communications uses interactive problem solving to help customers solve their own service problems.

AT&T is testing a Web interface that enables customers to add ports and circuits to their networks. Customers can also open trouble tickets and begin the repair process through the interface. Sprint's InTouch service provides customers with a customized real-time view of their companies' networks, available through the Sprint Web site [18]. Customers can review trouble tickets and see operational messages about network traffic.

Scorecard

For a technology industry, telecommunications providers are generally passive, not proactive, in using technology as a marketing weapon. MCI, whose market presence was launched through innovative marketing technology, is still in the forefront of an industry that should lead the world and does not. With the possible exception of BellSouth and GTE, whose Web sites offer a free e-mail address for users, telecommunications providers' corporate Web sites do not give users a compelling reason to return frequently.

Airlines use the Web to send weekly e-mails promoting weekend airfare specials and sell capacity that would go unused. Telecommunications providers could set up and advertise weekend rates to a limited customer base, and they have not yet done so. Again, this is probably not feasible when billing technologies are so unsophisticated. In some cases, the marketing opportunity is captured and lost. Telecommunications providers such as US West and Ameritech already advise customers on their Web sites whether desired features are available to them. While it is innovative for a local service provider to tell customers which services are available in their geographical area, it would demonstrate marketing initiative if the provider notified the curious customer, by e-mail or phone, of planned upgrades.

There are indications, though, that telecommunications service providers will rise to the occasion as the full range of providers participate in local and long-distance markets. Telecommunications service providers are already demonstrating a willingness to invest in new technology platforms and outsourced technology talent. Their expenditures in data warehouses and customer care systems present evidence of their eagerness to invest in systems that primarily benefit the marketing function. Upgrading legacy systems to meet interim regulatory requirements and year 2000 compliance is resource-consuming but necessary. When these investment requirements are fulfilled, telecommunications providers will redirect some of that expenditure to more innovative use. The competitive intensity and competitive parity already apparent in the industry will work to the advantage of the customer. Providers will innovate, copy, and surpass at a faster rate than in the past, or they will disappear. Marketing is the deciding factor.

SELF-ASSESSMENT—STRATEGIC TECHNOLOGY

The following questions will assist telecommunications marketers in analyzing whether strategic technology is an advantage for their company.

- Does your company support strategic marketing technologies?

- Is strategic technology a part of your company's marketing strategy? Is advanced technology necessary to compete in targeted markets?

- Are your company's mission-critical systems such as billing and customer care superior to those of competitors, average, or below the capabilities of competitors?

- If your company does not consider customer care to be an important differentiating factor, can low-cost technologies such as IVR and Web-based customer service reduce support costs?

- How will IP technologies affect the markets you presently serve? Does management have a plan to compete with or embrace IP network architecture?

References

[1] Caulfield, Brian, "Separating ISPs from the Telcos," *Internet World*, Vol. 4, No. 33, p. 27.

[2] Caulfield, Brian, "Business Slow to Adopt IP Telephony as Quality Lags," *Internet World*, Vol. 4, No. 36, p. 29.

[3] Lyford, Richard, "Are You Ready for Convergence Billing?," *Telephony*, Vol. 233, No. 7.

[4] Wilson, Tim, "The Buck Stops at the Help Desk—Carriers Turn Attention to Customer Service," *InternetWeek*, Issue 700, p. t19.

[5] Thyfault, Mary, "Clearer Connection with Customers," *InformationWeek*, Issue 700, pp. 245–248.

[6] Grambs, Peter, and Patrick Zerbib, "Iridium Customer Care," *Telephony*, Vol. 235, No. 19.

[7] Rennier, Steve, and Mark Steinmetz, "SBC Serves End Users with Ease," *Telephony*, Vol. 235, No. 5, pp. 34–38.

[8] Zieger, Anne, "Turn It On," *Telephony*, Vol. 235, No. 22, pp. 20–28.

[9] Wilder, Throop, and Stephen Van Beaver, "Mixed Nuts: Automated Provisioning, HFC, and IP," *America's Network*, Vol. 102, No. 17, pp. 30–34.

[10] Gewirtz, Rivka, "Interactive Voice Response Applications: More Than Just an Extra," *Telecom Business*, Vol. 1, No. 10, pp. 29–32.

[11] St. Ledger, Bob, "Solving Customer Care and Marketing Problems with Call Center Technology," *Telecommunications*, Americas Edition, Vol. 31, No. 7, p. 47.

[12] Shachtman, Noah, "Group Think—Employees are Shattering the Traditional Corporate Structure With Intranets," *InformationWeek*, Issue 684, pp. 77–84.

[13] Alexander, Steve, "Telecom Calls Intranet Shots," *Computerworld*, Vol. 31, Issue 34, pp. C2–C3.

[14] McGinity, Meg, "Online Billing Pays Off," *tele.com*, Vol. 3, No. 11, pp. 37–38.

[15] Wilson, Tim, "Ironically, Telcos Lag in E-Commerce Data," *InternetWeek*, February 17, 1998.

[16] Thyfault, Mary E., "Web Billing Breakthrough," *InformationWeek*, Issue 707, p. 26.

[17] Lawton, George, "Digital Facelift," *Global Telephony*, Vol. 6, Issue 7, p. 18.

[18] Bucholtz, Chris, "Looking Over Sprint's Shoulder," *Telephony*, Vol. 231, No. 13, p. 16.

Acronyms

ACD   automated call distribution

AOL   America Online

BT British Telecom

CAP   competitive access provider

CHAID   chi-squared automated interaction detector

CLEC   competitive local exchange carrier

CRM   customer relationship management

CTI   computer-telephony integration

DBS   direct broadcast satellite (television)

DSL   digital subscriber line

EDI electronic data interchange

ERP enterprise resource planning

FCC Federal Communications Commission (U.S.)

ICI Intermedia Communications, Inc.

IDC International Data Corporation

ILEC incumbent local exchange carrier

IP Internet protocol

ISDN integrated services digital network

ISP Internet service provider

IVR interactive voice response

IXC interexchange carrier

LEC local exchange carrier

MLM multilevel marketing

NARB National Advertising Review Board

NCR National Cash Register

OFTEL Office of Telecommunications (U.K.)

OSS operation support system

PC personal computer

PCS personal communications services

PEST political, economic, social, technological (analysis)

RAD rapid application development

RBOC regional Bell operating company

SBC Southwestern Bell Communications

SCIP   Society for Competitive Intelligence Professionals

SEC   Securities and Exchange Commission (U.S.)

SFA   sales force automation

SNET   Southern New England Telephone (acquired by SBC)

SOHO   small office/home office

SWOT   strengths, weaknesses, opportunities, threats (analysis)

TCG   Teleport Communications Group

TCI   Tele-Communications, Inc.

TMN   Telecommunications Management Network

VAR   value-added reseller

WAN   wide area network

WATS   wide area telephone service

About the Author

Karen G. Strouse is the owner of Management Solutions, a consulting firm specializing in the changing telecommunications industry. Her client list includes large and mid-sized telecommunications providers.

Ms. Strouse has assisted telecommunications service providers in developing entry strategies for new lines of business. She has conducted industry and competitive analyses in support of diversification efforts and assisted management in devising strategies, marketing approaches, pricing, and infrastructure to support new ventures. In addition, Ms. Strouse has helped clients to prepare business and strategic plans, develop acquisition strategy, perform competitive analysis, and improve operational effectiveness. In various industries, she has directed teams to improve business operations through reengineering of business processes and assisted management in constructing effective and efficient organizational structures.

Her experience includes a 13-year career with Bell Atlantic Corporation, during which she served in a variety of capacities, including director of strategic planning. After working with Bell Atlantic, she was a consulting manager with international firm Deloitte & Touche for five years. Ms. Strouse founded Management Solutions in 1992. Her e-mail address is kstrouse@msn.com.

Ms. Strouse holds B.A. and M.A. degrees in communications from Temple University and an M.B.A. in finance from St. Joseph University in Philadelphia.

Index

Recent Titles in the Artech House Telecommunications Library

Vinton G. Cerf, Senior Series Editor

Understanding Token Ring: Protocols and Standards, James T. Carlo, Robert D. Love, Michael S. Siegel, and Kenneth T. Wilson

Videoconferencing and Videotelephony: Technology and Standards, Second Edition, Richard Schaphorst

Visual Telephony, Edward A. Daly and Kathleen J. Hansell

World-Class Telecommunications Service Development, Ellen P. Ward

For further information on these and other Artech House titles, including previously considered out-of-print books now available through our In-Print-Forever® (IPF®) program, contact:

Artech House
685 Canton Street
Norwood, MA 02062
Phone: 781-769-9750
Fax: 781-769-6334
e-mail: artech@artechhouse.com

Artech House
46 Gillingham Street
London SW1V 1AH UK
Phone: +44 (0)171-973-8077
Fax: +44 (0)171-630-0166
e-mail: artech-uk@artechhouse.com

Find us on the World Wide Web at:
www.artechhouse.com